Fourier Series

© 2005 by

The Mathematical Association of America (Incorporated)

Library of Congress Catalog Control Number 2004113541

ISBN 0-88385-740-5

Printed in the United States of America

Current Printing (last digit):
10   9   8   7   6   5   4   3   2   1

Fourier Series

by

Rajendra Bhatia
Indian Statistical Institute

Published and Distributed by
THE MATHEMATICAL ASSOCIATION OF AMERICA

CLASSROOM RESOURCE MATERIALS

Classroom Resource Materials is intended to provide supplementary classroom material for students—laboratory exercises, projects, historical information, textbooks with unusual approaches for presenting mathematical ideas, career information, etc.

MAA Service Center
P. O. Box 91112
Washington, DC 20090-1112
1-800-331-1MAA
FAX: 1-301-206-9789
www.maa.org

Contents

Preface

These notes provide a quick and brief introduction to Fourier Series. The emphasis is not only on the mathematics but also on the history of the subject, its importance, its applications and its place in the rest of science.

I first learnt about Fourier Series as a student of physics. Together with several other assorted topics, they formed a ragbag course called Mathematical Physics from which, when the time came, *real* physics courses would pick what they wanted. A little later, as a student of mathematics I came across Fourier Series in the middle of a course on Mathematical Analysis. On each occasion, my teachers and my books (all good) managed to keep a secret which I learnt later. Fourier Series are not just tools for the physicist and examples for the mathematician. They are directly responsible for the development of nearly one half of mathematical analysis over the last two centuries.

These notes have been consciously designed to reveal this aspect of the subject and something more. The development of Fourier Series is illustrative of a recurrent pattern in modern science. I hope the reader will see this pattern emerge from our discussion.

This book can be used by a variety of students. Mathematics students at the third year undergraduate level should be able to follow *most* of the discussion. Typically, such students may have had their first course in Analysis (corresponding to Chapters 1–8 of *Principles of Mathematical Analysis* by W. Rudin) and have a good working knowledge of complex numbers and basic differential equations. Such students can learn about Fourier Series from this book and, at the same time, reinforce their understanding of the analysis topics mentioned above. More preparation is required for reading Chapter 4, a part of Section 2.5, Sections 5.2 and 5.4. These parts presuppose familiarity with Lebesgue spaces and elements of Functional Analysis usually taught in the fourth year of an undergraduate or the first year of a graduate program. At many places in the book the reader will see a statement like "*Let f be a continuous or, more generally, an integrable function.*" Here there is a choice. If it appears easier to handle continuous functions, the reader need not be worried about discontinuous ones at this stage.

Thus the material in this book can be used either to augment an Analysis course or to serve as the beginning of a special course leading to more advanced topics in Harmonic Analysis. It can also be used for a reading project. Some readers may be happy reading just Chapter 0 outlining the history of the subject; in some sense that captures the spirit of this book. Others may enjoy the several tidbits offered in Chapter 3.

Two editions of this book have appeared in India before this *Classroom Resource Materials* edition. I am much obliged to the editors of this series and to colleagues and friends H. Helson, A. I. Singh, S. K. Gupta, S. Serra and R. Horn for their comments and advice. The computer drawings were made by a former student S. Guha. I am thankful to him and to A. Shukla for preparing the electronic files.

0

A History of Fourier Series

A large part of mathematics has its roots in physics. Fourier series arose in the study of two simple physical problems—the motion of a vibrating string and heat conduction in solids. Attempts to understand what these series meant and in what sense they solved these two problems in physics have contributed to the origin and growth of most of modern analysis. Among ideas and theories that owe their existence to questions arising out of the study of Fourier series are Cantor's theory of infinite sets, the Riemann and the Lebesgue integrals and the summability of series. Even such a basic notion as that of a "function" was made rigorous because of these problems. A brief sketch of the history of Fourier series is given in this chapter. You could read it now and again at the end of this course when you will understand it better.

1. The motion of a vibrating string

It is a common experience that when an elastic string (such as a metal wire) tied at both ends is plucked or struck it begins vibrating. For simplicity, regard the string as a one-dimensional object occupying, when at rest, the interval $0 \leq x \leq 1$ in the x-y plane. At time $t = 0$ the string is displaced to its *initial position,* keeping its end points fixed, i.e., the shape of the string now becomes like the graph of a function

$$y = f(x) \qquad 0 \leq x \leq 1,$$

$$f(0) = f(1) = 0.$$

If the string is now released it starts vibrating. This is the model for a *plucked string.* We can also consider the *struck string* being described as: at time $t = 0$ each point x is imparted a velocity $g(x)$ in the y-direction. This is called the initial velocity. Once again the string will vibrate. Let $y = y(x, t)$ describe the graph of the displacement of the string at time t.

The motion is governed by the equations

(i) $y_{tt} = a^2 y_{xx}$,   *(the wave equation)*

(ii) $y(0, t) = y(1, t) = 0$ for all t,   *(boundary conditions)*

(iii) $y(x, 0) = f(x), y_t(x, 0) = g(x)$.   *(initial conditions)*

Here y_t and y_x denote the derivatives of y with respect to t and x. (See Section 5.3 for a derivation of the wave equation). What is the solution of this system?

One can see that for each integer k,

$$y(x, t) = \sin k\pi x \ \sin ak\pi t \tag{1}$$

and

$$y(x, t) = \sin k\pi x \ \cos ak\pi t, \tag{2}$$

both satisfy (i). They also satisfy (ii). The first satisfies (iii) if $f(x) = 0$ and $g(x) = ak\pi \sin k\pi x$. The second satisfies (iii) if $f(x) = \sin k\pi x$ and $g(x) = 0$.

Now note that any finite linear combination of functions like (1) and (2), i.e., a function

$$y(x, t) = \sum_k \sin k\pi x(\alpha_k \cos ak\pi t + \beta_k \sin ak\pi t), \tag{3}$$

also satisfies (i) and (ii). It satisfies (iii) for certain choices of f and g. But if f and g are any arbitrary functions, then obviously a finite sum of the form (3) is not likely to satisfy the conditions (iii). Now notice that an *infinite* sum of the form (3) will also satisfy (i) and (ii) *provided term by term differentiation of this series is permissible*. So the questions arise: When is such differentiation permissible? If this series solution converges and can be differentiated, can we now make it satisfy conditions (iii) for any preassigned f and g? Are there any other solutions for our problem?

Questions like this constitute the *heart of analysis*. It turns out that if f and g are sufficiently smooth, the solution we have outlined works, and is unique. (Functions in the vibrating string problem are piecewise C^2 and that is smooth enough.) Fourier series were introduced and developed in an attempt to answer such questions. And it is in the course of this development that such basic concepts of mathematics as *set* and *function* were made precise.

2. J. D'Alembert

The wave equation that describes the motion of a vibrating string was derived by J. D'Alembert in 1747. He gave a simple and elegant solution in the form

$$y(x, t) = v(at + x) - v(at - x),$$

which can be interpreted as a sum of two travelling waves, one moving to the left and the other to the right. (See Section 5.3, where we derive this solution via Fourier series. D'Alembert's solution was simpler; but the method of separation of variables we use in these notes is more general.)

When we are studying the motion of a plucked string: $y(x, 0) = f(x)$, $y_t(x, 0) = 0$, there is one variable f in the problem and the solution also involves one variable v which can be determined from f. So D'Alembert believed he had solved the problem completely.

3. L. Euler

At this time the word *function* had a very restricted meaning for mathematicians. A function meant a *formula* like $f(x) = x^2$, $f(x) = \sin x$, or $f(x) = x \tan x^2 + 1747e^x$. The formula could be complicated but it had to be a single analytic expression. Something like $f(x) = 3x^2$ when $0 \leq x \leq \frac{1}{2}$, and $f(x) = (1 - x^2)$ when $\frac{1}{2} \leq x \leq 1$ was not thought of as a function on $[0, 1]$. Euler thought that the initial position of the string need not always be a "function." The string could well be plucked to a shape as in Figure 1 and then released. Here different parts of the string are described by different functions. But Euler claimed that the travelling wave solution of D'Alembert would still be valid; now the solution would involve several different functions instead of a single function v.

In other words, at that time, *function* and *graph* meant two different things. Every function can be represented by its graph but not every "graph" that could be drawn was the graph of a function. Euler's point was that the initial displacement of the string could be any *graph* and the travelling wave solution should work here also, the two waves themselves being "graphs" now. This "physical" reasoning was rejected by D'Alembert; his "analytical" argument worked only for functions.

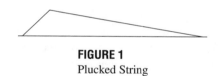

FIGURE 1
Plucked String

4. D. Bernoulli

In 1755, Daniel Bernoulli gave another solution for the problem in terms of *standing waves*. This is best understood by considering the solution

$$y(x, t) = \sin k\pi x \, \cos ak\pi t.$$

When $k = 1$, the points $x = 0$ and $x = 1$, i.e., the two end points of the string, remain fixed at all times, and all other points move, the motion of any point being given as a cosine function of time. When $k = 2$ the points $x = 0$, $x = 1$ and also the point $x = 1/2$ remain fixed at all times; the rest of the points all move; at any fixed time the string has the shape of a sine wave; and the motion of any point is given by a cosine function of time. The point $x = 1/2$ is called a *node*. For an arbitrary $k > 1$, the end points and the points $1/k, 2/k, \ldots, (k - 1)/k$ on the string remain fixed while all other points move as described. These are called "standing waves"; the (interior) fixed points are called *nodes* and the motion for $k = 1, 2, \ldots$ is called the *first harmonic*, the *second harmonic* and so on. Bernoulli asserted that every solution to the problem of the plucked string ($y(x, 0) = f(x)$, $y_t(x, 0) = 0$) is a sum of these harmonics, and the sum could be infinite.

Euler strongly objected to Bernoulli's claim. His objections were on two grounds. First of all, Bernoulli's claim would imply that *any* function $f(x)$ could be represented as

$$f(x) = \sum \alpha_k \sin k\pi x, \tag{4}$$

because at time $t = 0$ the initial position could be any $f(x)$. However, the right-hand side of (4) is a periodic function, whereas the left-hand side is completely arbitrary; further the right-hand side is an odd function of x whereas the left-hand side is completely arbitrary. Second, this would not, in any case, be a *general* solution to the problem. The right-hand side of (4) is an analytic formula, even though an infinite series, hence is a function. However, as he had pointed out earlier in connection with D'Alembert's solution, the initial position of the string could be any graph but not necessarily a function. So, Euler believed that D'Alembert's travelling wave solution was applicable to the general case of the initial position being a graph—which D'Alembert himself did not believe—but Bernoulli's solution would be applicable only to functions and to very restricted ones.

Bernoulli did not contest Euler's point about functions and graphs. However, he insisted that his solution was valid for all functions. His answer to Euler's first objection was that the series (4) involved infinitely many coefficients α_k and by choosing them properly one could make the series take the value $f(x)$ for infinitely many x. Of course, now we know that if two functions f and g are equal at infinitely many points, then they need not be equal at all points. However, at that time the nature of infinity and of a function was still not clearly understood and so this argument sounded plausible. Euler did not accept it, but for different reasons.

The series (4) is now known as a *Fourier series* and the coefficients α_k are called the *Fourier coefficients* of f. Now we know that not every continuous function can be written in this form, but still Bernoulli was almost right, as we will see. Assuming that f can be represented as (4) and that the series can be integrated term by term, the coefficients α_k are easily seen to be

$$\alpha_k = 2 \int_0^1 f(x) \sin k\pi x \, dx. \tag{5}$$

Euler derived this formula by a complicated argument and then noticed it was an easy consequence of the *orthogonality relations*

$$\int_0^1 \sin k\pi x \ \sin m\pi x \, dx = 0, \qquad k \neq m. \tag{6}$$

5. J. Fourier

In 1804, Joseph Fourier began his studies of the conduction of heat in solids and in three remarkably productive years discovered the basic equations of heat conduction, developed new methods to solve them, used his methods to analyse several practical problems and supplied experimental evidence to support his theory. His work was described in his book *The Analytical Theory of Heat*, one of the most important books in the history of physics.

Let us describe one of the simplest situations to which Fourier's analysis can be applied. Consider a 2-dimensional disk, say the unit disk in the plane. Suppose the temperature at each point of the boundary of the disk is known. Can we then find the temperature at any point inside the disk? At *steady state* the temperature $u(r, \theta)$ at a point (r, θ) obeys the equation

(i)
$$(ru_r)_r + 1/r \, u_{\theta\theta} = 0.$$

This is called the *Laplace equation in polar form.* (See Section 1.1 for the derivation of this equation from basic principles.) We are given the temperatures at the boundary, i.e., we know

(ii)
$$u(1, \theta) = f(\theta),$$

where f is a given continuous function. The problem is to find $u(r, \theta)$ for all (r, θ). This is the same kind of problem as we considered earlier for the vibrations of a string. Using an analysis very much like that of Bernoulli, Fourier observed that any finite sum

$$u(r, \theta) = \sum_{n=-N}^{N} A_n r^{|n|} e^{in\theta}$$

is a solution of (i). Of course, such a sum will not satisfy the *boundary condition* (ii) unless f is of a special type. Fourier asserted that an infinite sum

$$u(r, \theta) = \sum_{n=-\infty}^{\infty} A_n r^{|n|} e^{in\theta}$$

is also a solution of (i), and further by choosing A_n properly it can be made to satisfy the boundary condition (ii). In other words we can write

$$f(\theta) = \sum_{n=-\infty}^{\infty} A_n e^{in\theta} \tag{7}$$

when f is any continuous function.

So far, this is parallel to Bernoulli's analysis of the vibrating string. But Fourier went a step further and claimed that his method will work not only for f given by a single analytical formula but for f given by any *graph*. In other words, now there was to be no distinction between a function and a graph. Indeed, if Fourier's claim was valid, then every graph would also have a formula, namely the series associated with it.

Fourier, like Euler, calculated the coefficients A_n occurring in (7), by a laborious (and wrong) method. These are given by

$$A_n = \frac{1}{2\pi} \int_{-\pi}^{\pi} f(\theta) e^{-in\theta} d\theta. \tag{8}$$

The series (7) is now also called a *Fourier series* and the coefficients (8) the *Fourier coefficients* of f. Fourier noticed that the coefficients A_n are meaningful whenever f is a graph bounding a definite area (integrable in present day terminology). So he claimed his solution was valid for all such f.

Fourier's theory was criticised by Laplace and Lagrange, among others. However they recognised the importance of his work and awarded him a major prize.

Since $e^{in\theta} = \cos n\theta + i \sin n\theta$, the series (8) can be rearranged and written as

$$f(\theta) = \frac{a_0}{2} + \sum_{n=1}^{\infty}(a_n \cos n\theta + b_n \sin n\theta). \tag{9}$$

This too is called the Fourier series for f.

Now, what about Euler's two objections to Bernoulli's solution? (Euler died in 1783 and was not there to react to Fourier's work.) How could an arbitrary function be a sum of periodic functions? The way out of this difficulty is astonishingly simple. The given function is defined on some bounded interval, say $[0, 1]$ or $[0, \pi]$, where it represents some physical quantity of interest to us, like the displacement of a string. We can extend it outside this interval and make the extension periodic, and either odd or even. For example, if $f(x) = x, 0 \leq x \leq 1$, is the function given to us, we can define an extension of it by putting $f(x) = -x, -1 \leq x \leq 0$, and then further extending it by putting $f(x + 2k) = f(x)$, $-1 \leq x \leq 1, k \in Z$. This defines an even function of period 2. Its graph is shown in Figure 2.

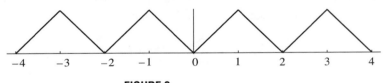

FIGURE 2
Even periodic extension of f

We could also have extended f from its original definition $f(x) = x, 0 \leq x \leq 1$, by putting $f(x) = x, -1 < x \leq 0$ and then extending it further by putting $f(x + 2k) = f(x), -1 < x \leq 1$, as before. Now we get an odd function of period 2; the graph of this function is shown in Figure 3. The Fourier series in the first case is

$$f(x) = \frac{1}{2} - \frac{4}{\pi^2} \cos \pi x - \frac{4}{9\pi^2} \cos 3\pi x - \cdots$$

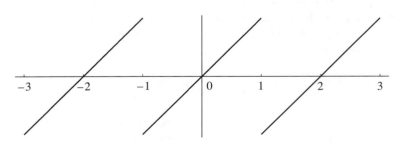

FIGURE 3
Odd periodic extension of f

and in the second case it is

$$f(x) = \frac{2}{\pi} \sin \pi x - \frac{2}{2\pi} \sin 2\pi x + \frac{2}{3\pi} \sin 3\pi x - \cdots$$

(These expansions, and several others are derived in Section 3.3). In the interval $[0, 1)$ both series converge to the same limit. Notice that the terms of one series are all even functions and of the other are all odd. Of course if a function is thought of as a formula, then these extensions are not functions. Fourier's idea of admitting functions which are identical on some interval but different elsewhere made it possible to apply his theory to a wide variety of situations. It also led to a critical examination of the notion of a function itself.

As Fourier observed, the coefficients A_n in (8) made sense for all functions for which the integral is finite. He believed that the series (7) would equal f for all such f at all θ. Here he was not right, but in a nontrivial sense. For all good (piecewise smooth) functions which occurred in his problems he was right.

6. P. Dirichlet

Fourier did not state or prove any statement about Fourier series that would be judged to be "correct" in a mathematics examination today. It was Dirichlet who took up Fourier's work and made it into rigorous mathematics. In the process he laid firm foundations for modern analysis.

First of all it was necessary to have a clear definition of a function. Dirichlet gave the definition which we learn now in our courses: a function is a rule which assigns a definite value $f(x)$ to any x in a certain set of points. Notice now that a function need no longer be a "graph," let alone a formula. In fact Dirichlet in 1828 gave an example that Fourier could not have imagined. This is the characteristic function of the set of rational numbers: $f(x) = 1$ if x is rational and $f(x) = 0$ if x is irrational. This function cannot be represented by any "graph." So now *the subject of analysis was no longer a part of geometry*. Notice also that this function does not "bound any area," so its Fourier coefficients cannot be calculated by Fourier's methods.

However, for all functions f that can be "drawn," i.e., for piecewise smooth functions f, Dirichlet *proved* that the Fourier series of f converges to $f(x)$ at every point x where f is continuous, and to the average value $\frac{1}{2}(f(x_+) + f(x_-))$ if f has a jump at x. Further, if f is smooth on an interval $[a, b]$, then its Fourier series converges *uniformly* to f on $[a, b]$. This was the first major convergence result for Fourier series. A later example is C. Jordan's theorem that says that continuous functions of bounded variation have convergent Fourier series. Dirichlet's theorem was a forerunner of several others giving sufficient conditions for a function to have a convergent Fourier series. (Some of these theorems are proved in Chapter 2, Section 3.)

7. B. Riemann

To handle functions that are not graphs, i.e., those that have more than finitely many turns and jumps, one would need to generalise the notion of an integral beyond the intuitive idea of the "area under a curve." Only then can we hope to calculate the Fourier coefficients of a function with infinitely many discontinuities. Riemann developed his theory of integration, which could handle such functions—that is the integral that we learn first now.

Using this integral Riemann gave an example of a function that does not satisfy Dirichlet's or Jordan's condition but has a pointwise convergent Fourier series. Riemann initiated the study of trigonometric series that need not be Fourier series of any function.

8. P. du Bois-Reymond

Dirichlet believed (and thought that he would soon prove) that the Fourier series of every continuous function, and perhaps of every Riemann integrable function, converges at every point. This belief came to be shared by other prominent mathematicians like Riemann, Weierstrass and Dedekind. In 1876, however, Du Bois-Reymond proved them wrong by constructing an example of a continuous function whose Fourier series is divergent at one point.

You might have come across the Weierstrass construction of an example of a continuous function which is not differentiable at any point. The function is obtained by successively constructing worse and worse functions. Du Bois-Reymond's example is a similar construction. This method is called the principle of *condensation of singularities*. (The construction of the example and the general method are described in Chapter 2, Section 5.)

9. G. Cantor

Cantor's theory of sets and infinite numbers is now the basis for all analysis. This theory too owes its existence, at least in part, to Cantor's interest in Fourier series. He observed that changing a function f at a few points does not change its Fourier coefficients. So the behaviour of f at a few points does not matter for Fourier analysis. How many points can be ignored in this way and what kind of sets do they constitute? This problem led Cantor to his study of infinite sets and cardinal numbers.

10. L. Fejér

Consider the series $\sum x_n$ where $x_n = (-1)^n$, $n \geq 0$. The partial sums s_n of this series are alternately 1 and 0. So the series does not converge. However, its divergence is different in character from that of the series $\sum n$ or $\sum 1/n$. In these latter cases the partial sums diverge to ∞, whereas in the former case the partial sums oscillate between 1 and 0 and hence *on an average* take the value $1/2$. We say that a series $\sum x_n$ is $(C, 1)$ summable or *Cesàro summable* if the sequence $\sigma_n = \frac{1}{n}(s_0 + s_1 + \cdots + s_{n-1})$, formed by taking the averages of the first n partial sums of the series, converges. Every convergent series is $(C, 1)$ summable; a series may be $(C, 1)$ summable but not convergent, for example this is the case when $x_n = (-1)^n$.

In 1904, Fejér proved that the Fourier series of every continuous function is $(C, 1)$ summable and, in this new sense, converges uniformly to the function. This theorem is extremely useful and gave an impetus to the study of summability of series. (Fejér's Theorem is proved in Chapter 2, Section 2.)

11. H. Lebesgue

It had already been noted, by Cantor among others, that changing a function at "a few" points does not alter its Fourier series, because the value of the integral (8) defining the Fourier coefficients is not affected. So it is not proper to ask whether the sum of the Fourier series of f is equal to f at *every* point; rather we should ask whether the two are equal everywhere except on those sets that are irrelevant in integration.

This problem led to a critical examination of the Riemann integral. Just as Riemann was motivated by problems in Fourier series to define a new concept of integration, the same motivation led Lebesgue to define a new integral that is more flexible. The notions of *sets of measure zero* and *almost everywhere equality of functions* now changed the meaning of function even more. We regard two functions f and g as identical if they differ only on a set of measure zero. Thus, Dirichlet's function (the characteristic function of the rationals) is equal to 0 "almost everywhere." So, from a *formula* to a *graph* to a *rule,* a function now became an *equivalence class*.

The Lebesgue integral is indispensable in analysis. Many basic spaces of functions are defined using this integral. We will talk about the L_p spaces

$$L_p = \left\{ f : \int |f|^p < \infty \right\}, \ 1 \le p < \infty,$$

in particular.

12. A.N. Kolmogorov

Before Du Bois-Reymond's example in 1876, mathematicians believed that the Fourier series of a continuous function will converge at every point. Since they had failed to prove this, the example made them think that perhaps the very opposite might be true—that there might exist a continuous function whose Fourier series diverges at *every* point.

In 1926 Kolmogorov proved something less but still very striking. He proved that there exists a Lebesgue integrable function defined on $[-\pi, \pi]$, i.e., a function in the space $L_1([-\pi, \pi])$, whose Fourier series diverges at every point.

13. L. Carleson

The function constructed by Kolmogorov is not continuous, not even Riemann integrable. However, after his example was published it was expected that sooner or later a continuous function with an everywhere divergent Fourier series would be discovered. There was a surprise once again. In 1966 Carleson proved that if f is in the space $L_2([-\pi, \pi])$, then its Fourier series converges to f at almost all points. In particular, this is true for continuous functions. So Fourier had been almost right!

Carleson's theorem is much harder to prove than any of the other results that we have mentioned. This whole investigation, which had occupied some of the very best mathematicians over two centuries, was nicely rounded off when in 1967, R.A. Hunt proved that the Fourier series of every function in L_p, $1 < p < \infty$, converges almost everywhere; and in the converse direction Y. Katznelson and J.P. Kahane proved in 1966 that given a set E

of Lebesgue measure 0 in $[-\pi, \pi]$, there exists a continuous function whose Fourier series diverges on E.

14. The L_2 theory and Hilbert spaces

We have already spoken of the space L_2, and earlier of orthogonality of trigonometric functions. The space L_2 is an example of a *Hilbert space*; in such spaces an abstract notion of distance and of orthogonality are defined. The functions $e_n(x) = e^{inx}/\sqrt{2\pi}$, $n = 0, \pm 1, \pm 2, \ldots$, constitute an *orthonormal basis* in the space $L_2([-\pi, \pi])$. So the Fourier series of f is now just an expansion of f with respect to this basis: $f(x) = \sum_{-\infty}^{\infty} A_n e_n(x)$; just as in the three-dimensional space $\mathbb{R}^3$, a vector can be written as a sum of its components in the three directions. The infinite expansion now is convergent in the metric of the space L_2, i.e.,

$$\lim_{N \to \infty} \left(\int_{-\pi}^{\pi} \left| f(x) - \sum_{n=-N}^{N} A_n e_n(x) \right|^2 dx \right) = 0.$$

This notion of convergence is different from that of convergence at every point, or at almost every point; but, in a sense, it is more natural because of the interpretation of the Fourier series as an *orthogonal expansion*. (We study this in Chapter 4.)

15. Some modern developments—I

One of the basic theorems in functional analysis says that the Hilbert spaces $L_2([-\pi, \pi])$ and l_2 are isomorphic. This theorem, called the Riesz–Fischer Theorem, is proved using Fourier series. It is taken as the starting point of J. von Neumann's "transformation theory" which he developed to show the equivalence of the two basic approaches to quantum mechanics called *matrix mechanics* and *wave mechanics*.

Another "transformation theory" achieving the same objective was developed by P.A.M. Dirac. In this theory a crucial role is played by the "δ-function." This has some unusual properties not consistent with classical analysis. (In von Neumann's book *Mathematical Foundations of Quantum Mechanics* the δ-function is described as "fiction.") Among other things the δ-function satisfies the properties $\delta(x) = \delta(-x)$, $\delta(x) = 0$ for all x except for $x = 0$ and $\int \delta(x) = 1$. Clearly now the integral could neither mean the area under a graph nor the Riemann integral nor the Lebesgue integral. The study of such objects, which are now called *generalised functions* or *distributions,* is a major branch of analysis developed by S.L. Sobolev and L. Schwartz. In some sense, spaces of such functions have become the most natural domain for Fourier analysis.

16. Some modern developments—II

We have talked only of Fourier series. A related object is the *Fourier integral* introduced by Fourier to study heat conduction in an infinite rod. This is the integral defined as

$$\hat{f}(t) = \int_{-\infty}^{\infty} f(x)e^{-itx}\,dx, \qquad t \in \mathbb{R}$$

and is called the *Fourier transform* of the function f. Notice the similarity between this and the Fourier coefficients where the variable t is replaced by the integer n, and the domain of integration is $[-\pi, \pi]$ instead of $(-\infty, \infty)$.

On the one hand this object has been used in such areas as heat conduction, optics, signal processing, and probability. On the other hand the subject has been made more abstract in a branch of mathematics called *harmonic analysis*. Integers and the interval $[-\pi, \pi)$ identified with the unit circle both form groups. Fourier series constitute harmonic analysis on these groups. The Fourier transform belongs to harmonic analysis on the group $\mathbb{R}$. A very similar analysis can be done on several other groups.

Major contributions to the "applied" branch were made by N. Wiener who developed what he called "generalised harmonic analysis." "Pure" harmonic analysis has developed into one of the central areas in mathematics. Some of the major figures in its development were E. Cartan, H. Weyl and Harish-Chandra. And this brings us to our final highlight.

17. Pure and applied mathematics

Fourier series were invented by Fourier who was studying a physical problem. It is no wonder then that they have applications. Of course not all creatures of "applied" mathematics have applications in as wide an area as Fourier series do. As this account shows, attempts to understand the behaviour of these series also laid down the foundations of rigorous analysis. Questions like uniform convergence, Cesàro summability and subjects like transfinite cardinals and Lebesgue measure are thought of as "pure" mathematics. Their history too is related to Fourier series.

Among the purest branches of mathematics is number theory—and surely it is a subject quite independent of Fourier series. Yet Hermann Weyl used Fejér's convergence theorem for Fourier series to prove a beautiful theorem in number theory, called the *Weyl Equidistribution Theorem*. Every real number x can be written as $x = [x] + \tilde{x}$, where $[x]$ is the *integral part* of x and is an integer, $\tilde{x}$ is the *fractional part* of x and is a real number lying in the interval $[0, 1)$. Weyl's theorem says that if x is an irrational number then, for large N, the fractional parts $\tilde{x}, (2x)^\sim, \ldots, (Nx)^\sim$ are scattered uniformly over $(0, 1)$. (This theorem is proved in Chapter 5, Section 1.)

One of the areas where Fourier series and transforms have major applications, is crystallography. In 1985 the Nobel Prize in Chemistry was given to H. A. Hauptman and J. Karle who developed a new method for calculating some crystallographic constants from their Fourier coefficients, which can be inferred from measurements. Two crucial ingredients of their analysis are Weyl's equidistribution theorem and theorems of Toeplitz on Fourier series of nonnegative functions.

Fast computers and computations have changed human life in the last few decades. One of the major tools in these computations is the Fast Fourier Transform introduced in 1965 by J. Cooley and J. Tukey in a short paper titled *An algorithm for the machine calculation of complex Fourier series*. Their idea reduced the number of arithmetic operations required in calculating a discretised version of the Fourier transform from $O(N^2)$ to $O(N \log N)$.

This went a long way in making many large calculations a practical job. (Note that if $N = 10^3$, then $N^2 = 10^6$ but $N \log N$ is approximately 7000.) In 1993 it was estimated that nearly half of all supercomputer central processing unit time was used in calculating the Fast Fourier Transform—used even for ordinary multiplication of large numbers.

Another major advance in the last few years is the introduction of *wavelets*. Here functions are expanded not in terms of Fourier series, but in terms of some other orthonormal bases that are suited to faster computations. This has led to new algorithms for signal processing and for numerical solutions of equations.

If conduction of heat is related to the theory of numbers, and if theorems about numbers are found useful in chemistry, this story has a moral. The boundary between deep and shallow may be sharper than that between pure and applied.

1

Heat Conduction and Fourier Series

In this chapter we derive the mathematical equations that describe the phenomenon of heat flow in a thin plate. We explain how, and in what sense, these equations are solved using Fourier Series. Our analysis leads to several questions. Some of these are answered in Chapters 1 and 2.

1.1 The Laplace equation in two dimensions

The equation

$$\frac{\partial^2 u}{\partial x^2} + \frac{\partial^2 u}{\partial y^2} = 0 \tag{1.1}$$

is called the 2-variable Laplace equation. Here $u(x, y)$ is a function on $\mathbb{R}^2$ which is of class C^2, i.e., u has continuous derivatives up to the second order.

This equation describes natural phenomena such as steady flow of heat in two dimensions. To understand this imagine a thin plate of some material. We can view this as a section of the plane with some boundary. If different parts of the plate are at different temperatures, then heat flows from a point at higher temperature to one at lower temperature according to *Newton's law of cooling*. This says that if Γ is a curve in the plane, then the rate at which heat flows across Γ is equal to $k \int_\Gamma (\partial u / \partial n) d\gamma$, where $u(x, y)$ is the temperature at the point (x, y) of the plate, $\partial u / \partial n$ denotes the derivative of u with respect to arc length along any curve normal to Γ and k is a constant called the *thermal conductivity* of the material of which the plate is made. (This is an experimental fact which describes the situation very accurately for moderate temperature gradients. We will accept this as a postulate.)

Now consider a small rectangular section of the plate

$$\{(x, y) : x_0 < x < x_1, y_0 < y < y_1\}.$$

If the plate is in thermal equilibrium, i.e., there is a steady state in which there is no net heat flow into this small section of the plate or out of it, then calculating the heat flow across the four boundaries of the above section according to Newton's law and equating it to 0 (steady state) we get

$$\int_{y_0}^{y_1} \left\{ \frac{\partial u}{\partial x}(x_1, y) - \frac{\partial u}{\partial x}(x_0, y) \right\} dy + \int_{x_0}^{x_1} \left\{ \frac{\partial u}{\partial y}(x, y_1) - \frac{\partial u}{\partial y}(x, y_0) \right\} dx = 0.$$

The negative signs here correspond to our understanding that if heat is flowing into the rectangle from one side, then it is flowing out of it from the opposite side. Divide this equation by $x_1 - x_0$ and then let $x_1 \to x_0$. One gets

$$\int_{y_0}^{y_1} \frac{\partial^2 u}{\partial x^2}(x_0, y) \, dy + \frac{\partial u}{\partial y}(x_0, y_1) - \frac{\partial u}{\partial y}(x_0, y_0) = 0.$$

Now divide by $y_1 - y_0$ and let $y_1 \to y_0$. This gives

$$\frac{\partial^2 u}{\partial x^2}(x_0, y_0) + \frac{\partial^2 u}{\partial y^2}(x_0, y_0) = 0.$$

This is the Laplace Equation (1.1) and as we have seen above it describes the phenomenon of steady state heat conduction in two dimensions.

We will find it more convenient to use polar coordinates. In these coordinates (1.1) can be transformed to

$$\frac{\partial^2 u}{\partial r^2} + \frac{1}{r} \frac{\partial u}{\partial r} + \frac{1}{r^2} \frac{\partial^2 u}{\partial \theta^2} = 0 \tag{1.2}$$

for $r \neq 0$, where as usual (r, θ) are the polar coordinates of a point in the plane.

Exercise 1.1.1. Derive (1.2) from (1.1) by a change of variables. Hint: Use the relations

$$\frac{\partial u}{\partial x} = \cos \theta \frac{\partial u}{\partial r} - \frac{1}{r} \sin \theta \frac{\partial u}{\partial \theta},$$

$$\frac{\partial u}{\partial y} = \sin \theta \frac{\partial u}{\partial r} + \frac{1}{r} \cos \theta \frac{\partial u}{\partial \theta},$$

which can be derived from the basic relations $x = r \cos \theta$, $y = r \sin \theta$.

Exercise 1.1.2. Derive (1.2) directly from Newton's law. Hints: Instead of the rectangular element that was used earlier consider now the element

$$\{(r, \theta) : r_0 < r < r_1, \theta_0 < \theta < \theta_1\}.$$

To calculate the rate of heat flow across the side $\theta = \theta_0$, choose as curves normal to it the circular curves $r = $ constant. Then note that the normal derivative with respect to the arc length along these curves is

$$\frac{\partial u}{\partial n} = \lim_{\delta\theta \to 0} \frac{u(r, \theta_0 + \delta\theta) - u(r, \theta_0)}{r \delta\theta} = \frac{1}{r} \frac{\partial u}{\partial \theta}(r, \theta_0).$$

Add the contributions from the four sides of the element and proceed as before.

Now consider a circular disk

$$D = \{(r, \theta) : 0 \le r \le 1, -\pi \le \theta \le \pi\}.$$

As usual the points $(r, -\pi)$ and (r, π) are considered to be identical. At steady state the temperature $u(r, \theta)$ for all points $0 < r < 1$ on D satisfies equation (1.2). Extend this to the origin by the natural condition

$$u(r, \theta) \text{ is continuous on } D. \tag{1.3}$$

We must also have

$$u(r, \theta) = u(r, \theta + 2\pi). \tag{1.4}$$

By a *boundary condition* we mean a condition of the type

$$u(1, \theta) = f(\theta), \tag{1.5}$$

where f is a given function defined on $[-\pi, \pi]$ that is continuous and for which $f(-\pi) = f(\pi)$.

The problem of finding a function u satisfying (1.2)–(1.5) is called a *Dirichlet problem*. This is a typical instance of a *boundary value problem* in physics. In the present situation this is: the temperature at the boundary of D is known to us and is given by (1.5); assuming that the temperature function inside D satisfies (1.2)–(1.4) can we find it? In the remaining part of this chapter we shall answer this question.

Exercise 1.1.3. Note that every constant function satisfies (1.2)–(1.4). Show that the function $u(r, \theta) = \log r$ satisfies (1.2) and (1.4) but not (1.3).

1.2 Solutions of the Laplace equation

Let us first derive solutions of two ordinary differential equations.

The first is the second order differential equation with constant coefficients:

$$y'' + ay' + by = 0. \tag{1.6}$$

The familiar method of solving this is the following. If L denotes the differential operator

$$L = \frac{d^2}{dx^2} + a\frac{d}{dx} + b,$$

then (1.6) can be written as

$$L(y) = 0. \tag{1.7}$$

Note that

$$L(e^{\alpha x}) = (\alpha^2 + a\alpha + b)e^{\alpha x},$$

and hence $y = e^{\alpha x}$ is a solution of (1.7) provided α is a root of the polynomial equation

$$p(\alpha) = \alpha^2 + a\alpha + b = 0. \tag{1.8}$$

This is called the *characteristic equation* associated with the differential equation (1.6). If this quadratic equation has two distinct roots α_1 and α_2, then the solutions $e^{\alpha_1 x}$ and $e^{\alpha_2 x}$ of (1.6) are linearly independent and hence their linear combination

$$y = c_1 e^{\alpha_1 x} + c_2 e^{\alpha_2 x} \tag{1.9}$$

gives the general solution of (1.6). If (1.8) has a double root α_0, then one solution of (1.6) is $e^{\alpha_0 x}$. In this case $p(\alpha_0) = p'(\alpha_0) = 0$ and one can see that

$$L(xe^{\alpha_0 x}) = p'(\alpha_0)e^{\alpha_0 x} + p(\alpha_0)xe^{\alpha_0 x} = 0.$$

So, in this case $xe^{\alpha_0 x}$ is another solution and the most general solution of (1.6) is of the type

$$y = c_1 e^{\alpha_0 x} + c_2 xe^{\alpha_0 x} . \tag{1.10}$$

The second equation that will arise in our analysis is the equation

$$x^2 y'' + axy' + by = 0, \qquad x > 0 \tag{1.11}$$

which is called an equation of *Cauchy type,* in which the coefficient of $y^{(n)}$ is x^n. Proceeding as earlier, let

$$L = x^2 \frac{d^2}{dx^2} + ax \frac{d}{dx} + b,$$

and note

$$L(x^\alpha) = p(\alpha)x^\alpha,$$

where,

$$p(\alpha) = \alpha(\alpha - 1) + a\alpha + b.$$

Hence $y = x^\alpha$ is a solution of (1.11) provided α is a solution of the characteristic equation

$$p(\alpha) = 0.$$

So, if this equation has two distinct roots α_1 and α_2, then the general solution of (1.11) is

$$y = c_1 x^{\alpha_1} + c_2 x^{\alpha_2}. \tag{1.12}$$

In case $p(\alpha) = 0$ has a double root α_0, then one can check that x^{α_0} and $x^{\alpha_0} \log x$ are two linearly independent solutions of (1.11) and hence the general solution, in this case, is

$$y = C_1 x^{\alpha_0} + C_2 x^{\alpha_0} \log x. \tag{1.13}$$

Now the Laplace equation can be solved by a standard device—the *separation method* which can (sometimes) reduce a partial differential equation to ordinary differential equations. This proceeds as follows. Let us assume, tentatively, that (1.2) has a solution of the form

$$u(r, \theta) = R(r)F(\theta), \tag{1.14}$$

where R and F are functions of r and θ alone, respectively. Then (1.2) becomes

$$r^2 R'' F + r R' F + R F'' = 0.$$

This is a mixture of two ordinary differential equations, which can be "separated" if we divide by RF and then rearrange terms as

$$r^2 \frac{R''}{R} + r \frac{R'}{R} = -\frac{F''}{F}.$$

Notice that now one side does not depend upon θ and the other side does not depend upon r. Hence the quantity on either side of this equation must be a constant. Denoting this constant by c, we get two equations

$$F'' + cF = 0, \tag{1.15}$$

$$r^2 R'' + r R' - cR = 0, \tag{1.16}$$

which are ordinary differential equations of the type (1.6) and (1.11), respectively.

Exercise 1.2.1. Show that the equations (1.15) and (1.16) have the following solutions:

(i) if $c > 0$, then

$$F(\theta) = A e^{i\sqrt{c}\theta} + B e^{-i\sqrt{c}\theta},$$

$$R(r) = a r^{\sqrt{c}} + b r^{-\sqrt{c}};$$

(ii) if $c = 0$, then

$$F(\theta) = A + B\theta,$$

$$R(r) = a + b \log r;$$

(iii) if $c < 0$, then

$$F(\theta) = A e^{\sqrt{-c}\theta} + B e^{-\sqrt{-c}\theta},$$

$$R(r) = a r^{i\sqrt{-c}} + b r^{-i\sqrt{-c}}.$$

Here A, B, a, b are constants.

(Though our function $u(r, \theta)$ is real, we will find it convenient to use complex quantities. If r is positive and α is any real number, then we have $r^{i\alpha} = e^{i\alpha \log r} = \cos(\alpha \log r) + i \sin(\alpha \log r)$).

In each of the three cases above $u = RF$ gives a solution of the Laplace equation (1.2). However, our other conditions rule out some of these solutions. The condition (1.4) says that $u(r, \theta)$ is a periodic function of θ with period 2π. The function $u = RF$ in case (iii) of Exercise 1.2.1 fails to meet this requirement. In case (ii) the function $F(\theta)$ is periodic only when $B = 0$, and the function $R(r)$ is bounded only when $b = 0$. Thus the only admissible solution in case (ii) is $u = Aa$. In case (i) the condition (1.4) is satisfied iff $c = n^2, n = 1, 2, \ldots$. With these values of c, R is bounded at the origin if $b = 0$. So the solution in Case (i) has the form $u = ar^n(Ae^{in\theta} + Be^{-in\theta}), n = 1, 2, \ldots$.

Thus, we have shown that all solutions of (1.2)–(1.4) that are of the form (1.14) are given by

$$u_n(r, \theta) = ar^n(Ae^{in\theta} + Be^{-in\theta}), \qquad n = 0, 1, 2, \ldots$$

where a, A, B are constants. Now notice that (finite) linear combinations of solutions of (1.2)–(1.4) are again solutions. So the sums

$$\sum_{n=-N}^{N} A_n r^{|n|} e^{in\theta}, \tag{1.17}$$

where A_n are constants all satisfy (1.2)–(1.4) for $N = 1, 2, \ldots$.

The constants A_n will be determined from the boundary condition (1.5). However it is clear that we cannot expect an arbitrary function f to be equal to a sum $\sum_{n=-N}^{N} A_n e^{in\theta}$.

One may now wonder whether an infinite sum

$$\sum_{n=-\infty}^{\infty} A_n r^{|n|} e^{in\theta} \tag{1.18}$$

is also a solution of (1.2)–(1.4). If so, then can the boundary condition (1.5) be satisfied by choosing the coefficients A_n properly, i.e., do we have

$$f(\theta) = u(1, \theta) = \sum_{n=-\infty}^{\infty} A_n e^{in\theta}. \tag{1.19}$$

Fourier asserted that this is indeed so for (virtually) every function f. He was not quite right. However, his analysis of this problem led to several developments in mathematics which we mentioned at the beginning.

To sum up: our analysis till (1.17) has been quite rigorous. Now we would like to answer the following questions:

1. For what functions f can we choose constants A_n so that the equation (1.19) is satisfied? If this equation is not satisfied in the usual sense of "pointwise convergence" is it true in some other sense? How are the coefficients A_n determined from f?

2. If A_n are chosen properly then is (1.18) indeed a solution of the Dirichlet problem?

Exercise 1.2.2. Some mystery may be taken out of the above problem if you use your knowledge of complex analysis. Any function $u(x, y)$ satisfying (1.1) is called a *harmonic* *function.* A function that is harmonic in the region D is the real part of a complex analytic function, i.e., there exists an analytic function $g(z)$ on D such that $g = u + iv$. Every such g has a power series expansion $\sum_{n=0}^{\infty} a_n z^n$ convergent in the interior of D. The coefficients a_n can be determined from g by Cauchy's integral formula. Recall that this power series does not converge at all points on the boundary of D. Use these facts to answer questions about Fourier series raised above.

1.3 The complete solution of the Laplace equation

We denote by T the boundary of the disk D. We can identify T with the interval $[-\pi, \pi]$, where the points $-\pi$ and π are regarded as the same. By a function on T we mean a function on $[-\pi, \pi]$ such that $f(-\pi) = f(\pi)$. Such a function can also be thought of as representing a periodic function on $\mathbb{R}$ with period 2π. So, we will use the terms "function on T" and "periodic function on $\mathbb{R}$ with period 2π" interchangeably.

Let f be a continuous function on T. We want to know whether it is possible to write f as an infinite series

$$f(\theta) = \sum_{n=-\infty}^{\infty} A_n e^{in\theta} \tag{1.20}$$

for some coefficients A_n depending on f. Suppose this is possible. Then what should A_n be? Recall that

$$\int_{-\pi}^{\pi} e^{im\theta}\, d\theta = \begin{cases} 0 & \text{if } m \neq 0, \\ 2\pi & \text{if } m = 0. \end{cases}$$

So, *if* we do have an expansion like (1.20) and *if* further this series could be integrated term by term, then we must have

$$A_n = \frac{1}{2\pi} \int_{-\pi}^{\pi} f(\theta) e^{-in\theta}\, d\theta. \tag{1.21}$$

So given a continuous, or more generally, an integrable function f on T let us define for each integer n,

$$\hat{f}(n) = \frac{1}{2\pi} \int_{-\pi}^{\pi} f(\theta) e^{-in\theta}\, d\theta. \tag{1.22}$$

Each $\hat{f}(n)$ is well defined. This is called the nth *Fourier coefficient* of f. The series

$$\sum_{n=-\infty}^{\infty} \hat{f}(n) e^{in\theta} \tag{1.23}$$

is called the *Fourier series* of f. At this moment we do not know whether this series converges, and if it converges whether at each point θ it equals $f(\theta)$.

Now let us return to the Dirichlet problem. Consider the series (1.18) where A_n are given by (1.21); we get a series

$$\sum_{-\infty}^{\infty} \int_{-\pi}^{\pi} \frac{1}{2\pi} r^{|n|} e^{in(\theta-t)} f(t)\, dt. \tag{1.24}$$

Now for $r < 1$ the series $\sum_{-\infty}^{\infty} r^{|n|} e^{in(\theta-t)}$ converges uniformly in t. (Use the M-test). Hence the sum and the integral in (1.24) can be interchanged. Also, using the identities

$$\sum_{n=0}^{\infty} x^n = \frac{1}{1-x},$$

$$\sum_{n=-\infty}^{-1} x^{-n} = \sum_{n=1}^{\infty} x^n = \frac{x}{1-x},$$

both valid when $|x| < 1$, one obtains

$$\sum_{-\infty}^{\infty} r^{|n|} e^{in(\theta-t)} = \frac{1-r^2}{1 - 2r\cos(\theta - t) + r^2} \quad \text{for} \quad r < 1.$$

Thus the series (1.24) converges for all r and θ, when $0 \le r < 1$ and $-\pi \le \theta \le \pi$. Call the sum of this series $u(r, \theta)$; we have

$$u(r, \theta) = \frac{1}{2\pi} \int_{-\pi}^{\pi} \frac{1-r^2}{1 - 2r\cos(\theta - t) + r^2} f(t)\, dt. \tag{1.25}$$

This is called the *Poisson integral* of f. The function

$$P(r, \theta) = \frac{1}{2\pi} \frac{1-r^2}{1 - 2r\cos\theta + r^2}, 0 \le r < 1, \tag{1.26}$$

is called the *Poisson kernel*.

A simple calculation shows that

$$\left(r^2 \frac{\partial^2}{\partial r^2} + r\frac{\partial}{\partial r} + \frac{\partial^2}{\partial \theta^2} \right) (r^n \cos n\theta) = 0, \quad \text{for all } n.$$

Since for $r < 1$ the series $\sum_{-\infty}^{\infty} r^{|n|} e^{in\theta}$, being absolutely and uniformly convergent, can be differentiated term by term, this shows that $u(r, \theta)$ as defined above is a solution of the Laplace equation. Thus $u(r, \theta)$ defined by (1.25) satisfies (1.2), (1.3) and (1.4). We will show that it also "satisfies" the boundary condition (1.5) in the sense that $u(r, \theta)$ tends to $f(\theta)$ uniformly as $r \to 1$.

Let f and g be two periodic functions on $\mathbb{R}$ with period 2π. If f and g are continuous, we define their *convolution* $f * g$ by

$$(f * g)(x) = \int_{-\pi}^{\pi} f(x - t)g(t)\, dt.$$

More generally, if f and g are two integrable functions, then

$$\int_{-\pi}^{\pi} |f(x-t)g(t)|\, dt < \infty$$

for almost all x. For these x we define $(f * g)(x) = \int_{-\pi}^{\pi} f(x-t)g(t)\, dt$.

It is easy to see that $f * g = g * f$.

Convolution is thus a commutative binary operation on the space of continuous periodic functions. Show that this operation is associative.

Does there exist an "identity" element for this operation? There is no continuous function g such that $g * f = f$ for all f. (See Exercise 4.2.5.) However, there is a modified notion of "approximate identity" in this context which is useful. We can find a family of functions Q_n such that $Q_n * f$ converges to f uniformly on $[-\pi, \pi]$ for all continuous functions f.

We call a sequence of functions Q_n on $[-\pi, \pi]$ a *Dirac sequence* if

(i) $Q_n(t) \geq 0,$ (1.27)

(ii) $Q_n(-t) = Q_n(t),$ (1.28)

(iii) $\displaystyle\int_{-\pi}^{\pi} Q_n(t)\, dt = 1,$ (1.29)

(iv)   for each $\varepsilon > 0$ and $\delta > 0$, there exists N such that for all $n \geq N$,

$$\int_{-\pi}^{-\delta} Q_n(t)\, dt + \int_{\delta}^{\pi} Q_n(t)\, dt < \varepsilon. \qquad (1.30)$$

You should ponder here a bit to see what these conditions mean. The first three conditions say that each Q_n is positive, even, and normalised so that its integral is 1. The last condition says that as n becomes large the graph of Q_n peaks more and more sharply at 0.

Theorem 1.3.1. *Let f be a continuous function on $[-\pi, \pi]$ and let Q_n be a Dirac sequence. Then $Q_n * f$ converges to f uniformly on $[-\pi, \pi]$ as $n \to \infty$.*

Proof Let $h_n = Q_n * f$. Using (1.29) we have

$$h_n(x) - f(x) = \int_{-\pi}^{\pi} [f(x-t) - f(x)]Q_n(t)\, dt.$$

Let ε be any given positive real number. The function f is uniformly continuous on $[-\pi, \pi]$; i.e., there exists a $\delta > 0$ such that

$$|f(x-t) - f(x)| < \frac{\varepsilon}{2} \quad \text{whenever} \quad |t| < \delta.$$

Let $M = \sup_{-\pi \leq x \leq \pi} |f(x)|$. Using (1.30) we can choose N such that for all $n \geq N$,

$$\int_{-\pi}^{-\delta} Q_n(t)\, dt + \int_{\delta}^{\pi} Q_n(t)\, dt < \frac{\varepsilon}{4M}.$$

Hence

$$\left(\int_{-\pi}^{-\delta} + \int_{\delta}^{\pi}\right) |f(x-t) - f(x)| Q_n(t)\, dt < 2M \frac{\varepsilon}{4M} = \frac{\varepsilon}{2}.$$

Also,

$$\int_{-\delta}^{\delta} |f(x-t) - f(x)| Q_n(t)\, dt < \int_{-\delta}^{\delta} \frac{\varepsilon}{2} Q_n(t)\, dt \le \frac{\varepsilon}{2}.$$

Hence

$$|h_n(x) - f(x)| < \varepsilon \quad \text{for all } n \ge N. \qquad \blacksquare$$

Later on, you might study the theory of distributions (or generalised functions) where you will come across Dirac's δ- function, which is not a function in the sense we understand it now, and which serves as an identity for the convolution operation.

In the same way, we define a *Dirac family* as a family of functions Q_r, $0 \le r \le 1$, where each Q_r satisfies the conditions (1.27), (1.28) and (1.29) and further for each ε and $\delta > 0$, there exists r_0 such that for $r \ge r_0$,

$$\int_{-\pi}^{-\delta} Q_r(t)\, dt + \int_{\delta}^{\pi} Q_r(t)\, dt < \varepsilon. \qquad (1.31)$$

In this case $Q_r * f$ converges to f uniformly as $r \to 1$.

Exercise 1.3.2. Show that the Poisson Kernel defined by (1.26) has the following properties:

 (i) $P(r, \varphi) \ge 0$,

 (ii) $P(r, \varphi) = P(r, -\varphi)$,

(iii) $P(r, \varphi)$ is a monotonically decreasing function of φ in $[0, \pi]$,

(iv) $\max_{-\pi \le \varphi \le \pi} P(r, \varphi) = P(r, 0) = (1 + r)/2\pi(1 - r)$,

 (v) $\min_{-\pi \le \varphi \le \pi} P(r, \varphi) = P(r, \pi) = (1 - r)/2\pi(1 + r)$,

(vi) $\int_{-\pi}^{\pi} P(r, \varphi) d\varphi = 1$,

(vii) for each $\varepsilon, \delta > 0$, there exists $0 < r_0 < 1$ such that, for $r_0 < r < 1$, we have

$$\int_{-\pi}^{-\delta} P(r, \varphi) d\varphi + \int_{\delta}^{\pi} P(r, \varphi) d\varphi < \varepsilon.$$

Hint: To prove (vii) proceed as follows. Since $P(r, \varphi)$ is an even function of φ you need to prove $\int_{\delta}^{\pi} P(r, \varphi) d\varphi < \varepsilon/2$. Since $P(r, \varphi)$ is monotonically decreasing,

$$\max_{\delta \le \varphi \le \pi} P(r, \varphi) = \frac{1}{2\pi} \frac{1 - r^2}{1 - 2r \cos \delta + r^2}.$$

Now for a fixed $\delta > 0$,

$$\lim_{r \to 1} \frac{1 - r^2}{1 - 2r \cos \delta + r^2} = 0.$$

Use this to choose the r_0 that is required.

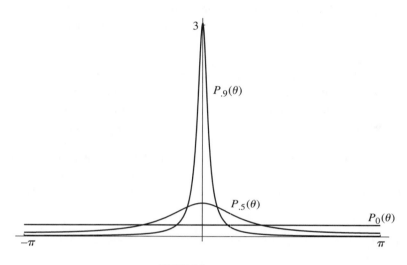

FIGURE 1
The Poisson kernel

Theorem 1.3.3. (Poisson's Theorem) *Let f be a continuous function on T and let $u(r, \theta)$ be defined by (1.25). Then as $r \to 1$, $u(r, \theta) \to f(\theta)$ uniformly.*

Proof For $0 \le r < 1$, let $P_r(\varphi) = P(r, \varphi)$ denote the Poisson kernel. Then

$$u(r, \theta) = (P_r * f)(\theta).$$

The family P_r is a Dirac family (by the preceding exercise). Use Theorem 1.3.1 now. ∎

Let us sum up what we have achieved so far. We have shown that $u(r, \theta)$ defined by (1.25) satisfies the conditions (1.2), (1.3) and (1.4) of the Dirichlet problem and further (for every continuous function f) it satisfies the boundary condition (1.5) in that $\lim_{r \to 1} u(r, \theta) = f(\theta)$ and the convergence is uniform in θ.

So we have found a solution to the Dirichlet problem. We will now show that the solution is unique, i.e., if $v(r, \theta)$ is another function that satisfies (1.2), (1.3) and (1.4) and if $v(r, \theta)$ converges to $f(\theta)$ uniformly as $r \to 1$, then $v(r, \theta) = u(r, \theta)$.

Exercise 1.3.4. Let f and g be continuous functions on T. Show that if $\hat{f}(n) = \hat{g}(n)$ for all integers n, then $f = g$. (Hint: Use Poisson's Theorem).

Exercise 1.3.5. Let f be a continuous function on T. Suppose the Fourier series of f converges uniformly. Show that its limit must be f. (Hint: Calculate the Fourier coefficients of the limit function.)

Warning: We have *not* proved that the Fourier series of a continuous function on T converges uniformly, or even pointwise, to f. In fact, this is false: there exists a continuous function whose Fourier series diverges on an uncountable set. However, we have shown that for each $0 \leq r < 1$, the series $\sum_{n=-\infty}^{\infty} r^{|n|} \hat{f}(n) e^{in\theta}$ converges and as r approaches 1 it converges uniformly to f.

Exercise 1.3.6. Let f be a continuous function on T and suppose the series $\sum_{n=-\infty}^{\infty} |\hat{f}(n)|$ converges. Show that the Fourier series of f converges to f uniformly on T.

This gives a sufficient condition for the convergence of Fourier series. Much stronger results will be proved in Chapter 2.

Theorem 1.3.7. (Uniqueness of the solution to Dirichlet's problem) *Let f be a continuous periodic function with period 2π. Let $u(r, \theta)$ be a function that satisfies* (1.2), (1.3) *and* (1.4) *and let $u(r, \theta)$ converge to $f(\theta)$ uniformly as $r \rightarrow 1$. Then u must be given by* (1.25).

Proof For a fixed $r < 1$, let $A_n(r)$ denote the Fourier coefficients of $u(r, \theta)$, i.e.,

$$A_n(r) = \frac{1}{2\pi} \int_{-\pi}^{\pi} u(r, \theta) e^{-in\theta} \, d\theta.$$

Differentiate with respect to r:

$$A_n'(r) = \frac{1}{2\pi} \int_{-\pi}^{\pi} u_r e^{-in\theta} \, d\theta,$$

$$(r A_n'(r))' = \frac{1}{2\pi} \int_{-\pi}^{\pi} (r u_r)_r e^{-in\theta} \, d\theta$$

$$= -\frac{1}{2\pi} \int_{-\pi}^{\pi} \frac{1}{r} u_{\theta\theta} e^{-in\theta} \, d\theta,$$

since u satisfies (1.2). Here u_r and u_θ denote the derivatives of u with respect to r and θ. Now integrate by parts and use (1.4) to get

$$(r A_n'(r))' = \frac{n^2}{r} A_n(r).$$

i.e.,

$$r^2 A_n''(r) + r A_n'(r) - n^2 A_n(r) = 0. \tag{1.32}$$

Now this is exactly the Cauchy type differential equation which we have seen earlier as (1.11). For $n \neq 0$ its solution is

$$A_n(r) = C_n r^{|n|} + D_n r^{-|n|}$$

where C_n and D_n are constants. Since

$$|A_n(r)| \leq \sup |u(r, \theta)|$$

and $u(r, \theta)$ is bounded on D we must have

$$A_n(r) = C_n r^{|n|}, \qquad n \neq 0.$$

The C_n are determined as follows:

$$C_n r^{|n|} = A_n(r) = \frac{1}{2\pi} \int_{-\pi}^{\pi} u(r, \theta) e^{-in\theta} \, d\theta.$$

Letting $r \to 1$ this gives

$$C_n = \frac{1}{2\pi} \int_{-\pi}^{\pi} f(\theta) e^{-in\theta} \, d\theta = \hat{f}(n), n \neq 0.$$

Thus $A_n(r) = \hat{f}(n) r^{|n|}$, $n \neq 0$. For $n = 0$ the only bounded solution of (1.2) is

$$A_0(r) = \text{constant}.$$

Hence

$$\int_{-\pi}^{\pi} u(r, \theta) \, d\theta \text{ is independent of } r.$$

So for all $0 \leq r < 1$,

$$A_0(r) = A_0 = \frac{1}{2\pi} \int_{-\pi}^{\pi} u(r, \theta) \, d\theta = \frac{1}{2\pi} \int_{-\pi}^{\pi} f(\theta) \, d\theta.$$

We have shown that the Fourier coefficients of $u(r, \theta)$ are $\hat{f}(n) r^{|n|}$. But these are also the Fourier coefficients of (1.25). So the theorem follows from Exercise 1.3.4. ∎

Exercise 1.3.8.

(i) Prove the "mean value property" of temperature: at steady state the temperature at the centre of a disk is the average of the temperature on the boundary.

(ii) Prove the "maximum principle" and the "minimum principle": at steady state the hottest and the coldest points on a disk are at the boundary.

(iii) Can you relate these results to facts which you might have learnt in your Complex Analysis course?

Exercise 1.3.9.
Use Poisson's Theorem to prove the Weierstrass Approximation Theorem in the following form: let $f(\theta)$ be a continuous periodic function of period 2π, then f is a uniform limit of "trigonometric polynomials," i.e., finite sums of the form $\sum_{n=-N}^{N} a_n e^{in\theta}$.

Exercise 1.3.10.
Let f and g be two integrable functions on $[-\pi, \pi]$. Show that their convolution $f * g$ is also an integrable function. If either f or g is continuous, show that $f * g$ is continuous; and if either f or g is C^1, show that $f * g$ is also C^1.

Exercise 1.3.11. (An extension of Theorem 1.3.1 to integrable functions). Let f be integrable on $[-\pi, \pi]$ and let Q_n be a Dirac sequence. Show that

(i) $(Q_n * f)(x) \to f(x)$ if f is continuous at x;

(ii) if the left-hand and the right-hand limits of f at x exist, denote them by $f(x_-)$ and $f(x_+)$ respectively, and show $(Q_n * f)(x) \to \frac{1}{2}[f(x_+) + f(x_-)]$;

(iii) if f is continuous for each x in a closed interval I, then

$$(Q_n * f)(x) \to f(x) \text{ uniformly on } I.$$

2

Convergence of Fourier Series

We have defined the Fourier coefficients of f as

$$\hat{f}(n) = \frac{1}{2\pi} \int_{-\pi}^{\pi} f(\theta)e^{-in\theta} \, d\theta. \tag{2.1}$$

These are well defined for each continuous function on T, or more generally, for each integrable function on T. The Fourier series of f is the series

$$\sum_{n=-\infty}^{\infty} \hat{f}(n)e^{in\theta}. \tag{2.2}$$

Associated with the series is the sequence of its partial sums

$$S_N(f; \theta) = \sum_{n=-N}^{N} \hat{f}(n)e^{in\theta}, \tag{2.3}$$

$N = 0, 1, 2, \ldots$. If at a point θ of T the sequence (2.3) converges, we say that the Fourier series (2.2) converges at θ. It would have been nice if such convergence did take place at every point θ. Unfortunately, this is not the case. There are continuous functions f for which the series (2.2) diverges for uncountably many θ. Now we can proceed in two directions:

(1) Weaken the notion of convergence, or

(2) Strengthen the conditions on f.

In the first direction we will see that the Fourier series of every continuous function converges in the sense of *Abel summability* and *Cesàro summability*, both of which are weaker notions than pointwise convergence of the sequence (2.3). In the second direction we will see that if f is not only continuous but differentiable, then the series (2.2) is convergent at every point θ to the limit $f(\theta)$. More generally, this is true when f is not necessarily differentiable but is Lipschitz or is of bounded variation.

2.1 Abel summability and Cesàro summability

Consider any series

$$\sum_{n=1}^{\infty} x_n \tag{2.4}$$

with real or complex term x_n, $n = 1, 2, \ldots$. If for *every* real number $0 < r < 1$ the series $\sum_{n=1}^{\infty} r^n x_n$ converges and if $\sum_{n=1}^{\infty} r^n x_n$ approaches a limit L as $r \to 1$, then we say that the series (2.4) is *Abel summable* and its *Abel limit* is L.

Exercise 2.1.1.

(i) If the series (2.4) converges in the usual sense to L, show that it is also Abel summable to L.

(ii) Let x_n be alternately 1 and -1. Then the series (2.4) does not converge but is Abel summable and its Abel limit is $1/2$.

(iii) Let $x_n = (-1)^{n+1} n$. Show that the series (2.4) is Abel summable and its Abel limit is $1/4$.

(iv) In your Complex Analysis course you might have come across this notion while studying the radius of convergence of power series. (See, for example, "Abel's Limit Theorem" in *Complex Analysis* by L.V. Ahlfors.) Is there any connection between that theorem and what we are doing now?

Poisson's Theorem (Theorem 1.3.3) proved in Chapter 1 can now be stated as:

Theorem 2.1.2. *If f is a continuous function on T, then its Fourier series is Abel summable and has Abel limit $f(\theta)$ at every θ.*

Cesàro convergence is defined as follows. For the series (2.4) let

$$s_N = \sum_{n=1}^{N} x_n, \ N = 1, 2, \ldots$$

be the sequence of its partial sums. Consider the averages of these partial sums:

$$\sigma_n = \frac{s_1 + s_2 + \cdots + s_n}{n}.$$

If the sequence σ_n converges to a limit L as $n \to \infty$, we say that the series (2.4) is *summable to L in the sense of Cesàro,* or is *Cesàro summable* to L. This is sometimes also called $(C, 1)$ summability or *summability by the method of the first arithmetic mean.*

Exercise 2.1.3.

(i) If the series (2.4) converges in the usual sense to L, show that it is Cesàro summable to L.

(ii) If the series (2.4) is Cesàro summable to L, show that it is Abel summable to L.

(iii) Show that the converses of statements (i) and (ii) are false. (See the examples in Exercise 2.1.1.)

(iv) If $x_n \geq 0$, show that the series (2.4) is Cesàro summable if and only if it is convergent.

We will soon prove that if f is a continuous function on T, then its Fourier Series is Cesàro summable to the limit $f(\theta)$ for every θ. In view of the relationship between Cesàro convergence and Abel convergence (Exercise 2.1.3) this is a stronger result than Theorem 2.1.2. However, it is still fruitful to use Abel convergence because of its connection with the theory of complex analytic functions.

2.2 The Dirichlet and the Fejér kernels

For each integer n let $e_n(t) = e^{int}$. If f is a continuous function on T, then

$$(f * e_n)(\theta) = \int_{-\pi}^{\pi} f(t)e_n(\theta - t)\, dt$$

$$= e^{in\theta} \int_{-\pi}^{\pi} f(t)e^{-int}\, dt$$

$$= 2\pi \hat{f}(n)e^{in\theta}.$$

Hence, we can write the partial sums (2.3) as

$$S_N(f; \theta) = \frac{1}{2\pi} \sum_{n=-N}^{N} (f * e_n)(\theta) = (f * D_N)(\theta), \tag{2.5}$$

where

$$D_N(t) = \frac{1}{2\pi} \sum_{n=-N}^{N} e^{int}. \tag{2.6}$$

The expression $D_N(t)$ is called the *Dirichlet kernel*.

Exercise 2.2.1. Show that:

(i) $\displaystyle\sum_{n=1}^{N} e^{int} = \frac{\sin(Nt/2)}{\sin(t/2)} e^{i(N+1)t/2}, \quad$ for all $t \neq 2k\pi$. $\tag{2.7}$

(ii) $\displaystyle\left| \sum_{n=1}^{N} e^{int} \right| \leq \frac{1}{|\sin(t/2)|}, \quad$ for all $t \neq 2k\pi$.

(iii) $\displaystyle\sum_{n=1}^{N} \cos nt = -\frac{1}{2} + \frac{\sin((N + 1/2)t)}{2\sin(t/2)}, \quad$ for all $t \neq 2k\pi$.

(iv) $\displaystyle\sum_{n=1}^{N} e^{i(2n-1)t} = \frac{\sin Nt}{\sin t} e^{iNt}, \quad$ for all $t \neq k\pi$.

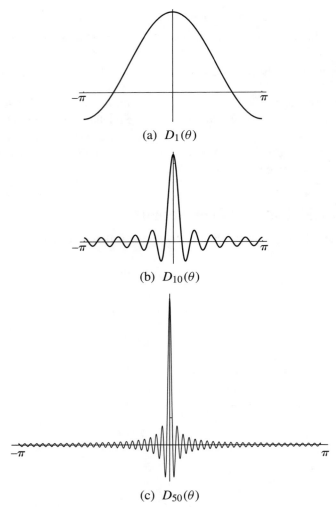

(a) $D_1(\theta)$

(b) $D_{10}(\theta)$

(c) $D_{50}(\theta)$

FIGURE 1
The Dirichlet kernel for different values of n

(v) $\displaystyle\sum_{n=1}^{N} \sin((2n-1)t) = \frac{\sin^2 Nt}{\sin t}$,   for all $t \neq k\pi$.

Hint: Write the sum in (2.7) as

$$e^{it}\frac{1-e^{iNt}}{1-e^{it}},$$

then use $1 - e^{it} = e^{it/2}(e^{-it/2} - e^{it/2})$.

Taking real parts of the two sides in (2.7) we get (iii).

Proposition 2.2.2. *The Dirichlet kernel has the properties*

(i) $D_N(-t) = D_N(t)$,

(ii) $\int_{-\pi}^{\pi} D_N(t)\,dt = 1$,

(iii) $D_N(t) = \dfrac{1}{2\pi}\dfrac{\sin(N + \frac{1}{2})t}{\sin t/2}$.

Proof Properties (i) and (ii) are obvious and (iii) follows from 2.2.1 (iii). ∎

The *Fejér kernel $F_n(t)$* is defined as

$$F_n(t) = \frac{1}{n}\sum_{k=0}^{n-1} D_k(t). \tag{2.8}$$

Exercise 2.2.3. Show that:

(i) $F_n(-t) = F_n(t)$.

(ii) $\int_{-\pi}^{\pi} F_n(t)\,dt = 1$.

(iii) $F_n(t) = \dfrac{1}{2n\pi}\dfrac{\sin^2 nt/2}{\sin^2 t/2} = \dfrac{1}{2n\pi}\dfrac{1-\cos\,nt}{1-\cos\,t}$.

Hint: Use Proposition 2.2.2 and Exercise 2.2.1 (v).

(iv) $F_n(t) \geq 0$ for all t.

(v) $F_n(t) \leq \dfrac{1}{n\pi(1 - \cos\,\delta)}$ for $0 < \delta \leq |t| \leq \pi$.

(vi) $F_n(t) = \displaystyle\sum_{j=-(n-1)}^{n-1}\left(1 - \frac{|j|}{n}\right)e^{ijt}$.

These properties of F_n can be used to conclude:

Theorem 2.2.4. *The sequence F_n is a Dirac sequence.*

This has a most important consequence:

Theorem 2.2.5. (Fejér's Theorem) *Let f be a continuous function on T. Then the Fourier series of f is Cesàro summable to f at every point of T. Further, the convergence of the sequence*

$$\sigma_n(f;\theta) = \frac{1}{n}\sum_{k=0}^{n-1} S_k(f;\theta)$$

to $f(\theta)$ is uniform on T.

Proof Note that $\sigma_n(f;\theta) = (f * F_n)(\theta)$. Use Theorem 1.3.1. ∎

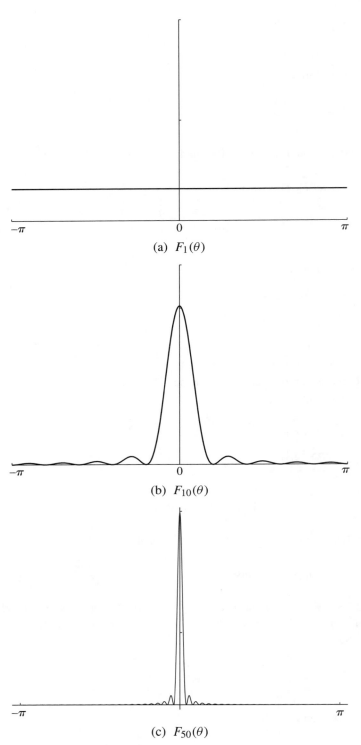

(a) $F_1(\theta)$

(b) $F_{10}(\theta)$

(c) $F_{50}(\theta)$

FIGURE 2
The Fejér kernel for different values of n

In (2.5) we observed that $S_N(f; \theta) = (f * D_N)(\theta)$. So *if* D_N were a Dirac sequence we could have concluded that the Fourier series of f converges to f. However, this is not the case. At this point we will do well to compare the three kernels $P_r(\theta)$, $D_n(\theta)$ and $F_n(\theta)$ of Poisson, Dirichlet and Fejér which we have come across. The following features should be noted:

1. $P_r(t)$ and $F_n(t)$ are nonnegative for all t but $D_n(t)$ is not.
2. All three are even functions, all attain a maximum at $t = 0$. But $P_r(t)$ decreases mono- tonically from 0 to π, whereas $D_n(t)$ and $F_n(t)$ are oscillatory.
3. For $D_n(t)$ and $F_n(t)$ the oscillations become more and more rapid as n increases. How- ever, $D_n(t)$ does not die out at $\pm\pi$, whereas $F_n(t)$ dies out at $\pm\pi$ for large n. $P_r(t)$ also approaches 0 for $t = \pm\pi$ as $r \to 1$.
4. The peak of all three at $t = 0$ goes to ∞ as $r \to 1$, or $n \to \infty$.
5. It is true that the integrals of all three over the interval $[-\pi, \pi]$ are 1. Since P_r and F_n are nonnegative we have $\int_{-\pi}^{\pi} |P_r(t)| \, dt = \int_{-\pi}^{\pi} |F_n(t)| \, dt = 1$. However, we will see that $\int_{-\pi}^{\pi} |D_n(t)| \, dt$ goes to ∞ with n.
6. Both D_n and F_n are polynomial expressions in e^{it}, i.e., they are finite linear combina- tions of e^{int}. Such expressions are called *exponential polynomials*. The Poisson kernel is not of this type.

The numbers

$$L_n = \int_{-\pi}^{\pi} |D_n(t)| \, dt = \frac{1}{2\pi} \int_{-\pi}^{\pi} \left| \frac{\sin(n + \frac{1}{2})t}{\sin t/2} \right| dt \tag{2.9}$$

are called the *Lebesgue constants*. We will see that as $n \to \infty$, $L_n \to \infty$ at the same rate as $\log n$.

Exercise 2.2.6. Show that

(i) $\sin t \le t$ for $0 \le t \le \pi$,
(ii) $t \le \dfrac{\pi}{2} \sin t$ for $0 \le t \le \dfrac{\pi}{2}$.

Exercise 2.2.7. Let $x_n = 1 + \frac{1}{2} + \cdots + \frac{1}{n} - \log n$. Show that x_n converges. The limit of x_n is called the Euler constant and is denoted by γ. Show that $0 < \gamma < 1$. Hints: Use the fact that $\int (dt/t) = \log t$.

(i) Show that

$$\sum_{k=2}^{n} \frac{1}{k} < \int_{1}^{n} \frac{dt}{t} < \sum_{k=1}^{n-1} \frac{1}{k}.$$

From this conclude that x_n is monotonically decreasing and $0 < x_n < 1$. So x_n con- verges to a limit γ.

(ii) Use the same idea to show

$$\frac{1}{n+1} < \log \frac{n+1}{n} < \frac{1}{n}.$$

Note for all $n > 1$ we can write

$$\log n = \log 2 + \log \frac{3}{2} + \cdots + \log \frac{n}{n-1}.$$

Use these two facts to obtain

$$1 - \log 2 < x_n < 1 - \left(\log 2 - \frac{1}{2}\right).$$

This shows $0 < \gamma < 1$. (It is not known whether γ is a rational number. An approximate value is $\gamma = 0.57722$. You are likely to come across this number when you study the gamma function.)

Exercise 2.2.8. The aim of this exercise is to show that L_n goes to ∞ like a constant multiple of $\log n$. This can be expressed in any of the following ways:

(a) There exist constants C_1 and C_2 such that

$$C_1 \log n < L_n < C_2 \log n \quad \text{for} \quad n \geq 2.$$

(b) The sequence $L_n - (4/\pi^2) \log n$ is bounded; i.e.,

(c) $L_n = \dfrac{4}{\pi^2} \log n + O(1)$.

Here is an outline of the proof:

(i) $L_n = \dfrac{2}{\pi} \displaystyle\int_0^{\frac{\pi}{2}} \left| \frac{\sin(2n+1)u}{\sin u} \right| du.$

(ii) By Exercise 2.2.6 (i)

$$L_n \geq \frac{2}{\pi} \int_0^{\frac{\pi}{2}} \left| \frac{\sin(2n+1)u}{u} \right| du.$$

(iii) Now write

$$\int_0^{\frac{\pi}{2}} \left| \frac{\sin(2n+1)u}{u} \right| du = \sum_{k=0}^{2n} \int_{\frac{k}{2n+1} \cdot \frac{\pi}{2}}^{\frac{k+1}{2n+1} \cdot \frac{\pi}{2}} \left| \frac{\sin(2n+1)u}{u} \right| du.$$

(iv) Each of these integrals can be estimated from below by replacing u occurring in the denominator by the larger quantity

$$\frac{k+1}{2n+1} \cdot \frac{\pi}{2}.$$

After this the resulting integrals are simple to evaluate.

(v) This gives

$$L_n \geq \frac{4}{\pi^2} \sum_{k=0}^{2n} \frac{1}{k+1}.$$

Use Exercise 2.2.7 now.

(vi) At step (ii) above use the inequality of Exercise 2.2.6 (ii) instead, to estimate L_n from above. Make a similar change in step (iv). Get an upper bound for L_n.

Exercise 2.2.9. Estimate L_n in another way as follows:

(i) $L_n = \dfrac{2}{\pi} \displaystyle\int_0^\pi \left| \dfrac{\sin(n + \frac{1}{2})u}{2 \sin \frac{u}{2}} \right| du.$

(ii) Split the integral in two parts; one from 0 to $1/n$ and the other from $1/n$ to π. By Exercise 2.2.1 (iii)

$$\left| \frac{\sin(n + \frac{1}{2})u}{2 \sin \frac{u}{2}} \right| \leq n + \frac{1}{2}.$$

So

$$\int_0^{1/n} \left| \frac{\sin(n + \frac{1}{2})u}{2 \sin \frac{u}{2}} \right| du \leq 1 + \frac{1}{2n}.$$

(iii) For the other integral, use Exercise 2.2.6 (ii) to bound the denominator of the integrand, and replace the numerator by 1. This gives

$$\int_{1/n}^\pi \left| \frac{\sin(n + \frac{1}{2})u}{2 \sin \frac{u}{2}} \right| du \leq \frac{\pi}{2} \int_{1/n}^\pi \frac{du}{u} = \frac{\pi}{2}(\log \pi + \log n).$$

(iv) So

$$L_n \leq \log n + \log \pi + \frac{2}{\pi}\left(1 + \frac{1}{2n}\right).$$

(v) Write this in the form

$$L_n \leq C \log n \quad \text{for} \quad n \geq 2.$$

2.3 Pointwise convergence of Fourier series

By the Weierstrass approximation theorem every continuous function on T is a uniform limit of exponential polynomials. (See Exercise 1.3.9; this follows also from Theorem 2.2.5). In other words if f is a continuous function on T, then for every $\varepsilon > 0$ there exists an exponential polynomial

$$p_N(t) = \sum_{n=-N}^{N} a_n e^{int} \tag{2.10}$$

such that

$$\sup_{-\pi \leq t \leq \pi} |f(t) - p_N(t)| \leq \varepsilon. \tag{2.11}$$

This can be used to prove:

Theorem 2.3.1. (The Riemann–Lebesgue Lemma) *If f is a continuous function on T, then*

$$\lim_{|n| \to \infty} \hat{f}(n) = 0.$$

Proof We want to show that given an $\varepsilon > 0$ we can find an N such that for all $|n| > N$ we have $|\hat{f}(n)| < \varepsilon$. Choose p_N to satisfy (2.10) and (2.11). Note that for $|n| > N$, $\hat{p}(n) = 0$. Hence for $|n| > N$ we have

$$\hat{f}(n) = \hat{f}(n) - \hat{p}(n) = (\widehat{f - p})(n).$$

But from (2.10) we get

$$|(\widehat{f - p})(n)| = \frac{1}{2\pi} \left| \int_{-\pi}^{\pi} [f(t) - p(t)]e^{-int} \, dt \right|$$

$$\leq \frac{1}{2\pi} \varepsilon . 2\pi = \varepsilon. \qquad \blacksquare$$

Exercise 2.3.2. Let f be a continuous function on T. Show that

$$\lim_{n \to \infty} \int_{-\pi}^{\pi} f(t) \cos nt = 0, \qquad \lim_{n \to \infty} \int_{-\pi}^{\pi} f(t) \sin nt = 0.$$

Exercise 2.3.3. Let f be a continuous function on T. Show that

$$\lim_{n \to \infty} \int_{-\pi}^{\pi} f(t) \sin\left(\left(n + \frac{1}{2}\right)t\right) dt = 0.$$

Hint: Use the two statements of Exercise 2.3.2 replacing $f(t)$ by $f(t) \sin \frac{t}{2}$ and $f(t) \cos \frac{t}{2}$ respectively.

Exercise 2.3.4. If f is continuously differentiable on $[a, b]$, use integration by parts to show that

$$\lim_{x \to \infty} \int_{a}^{b} f(t) \sin tx \, dt = 0.$$

This gives another proof of the Riemann–Lebesgue Lemma for such functions. More generally, use this method to prove this lemma for piecewise C^1 functions. [Definition: A function f on $[a, b]$ is called *piecewise continuous* if there exist a finite number of points

$a_j, a = a_0 < a_1 < \cdots < a_n = b$, such that f is continuous on each interval (a_j, a_{j+1}) and the limits $f(a_j+)$ and $f(a_j-)$ exist for all j. Such a function is called *piecewise* C^1, if the derivative f' exists and is continuous on each interval (a_j, a_{j+1}) and the limits $f'(a_j+)$ and $f'(a_j-)$ exist for all j.]

Exercise 2.3.5. Note that the Fourier coefficients $\hat{f}(n)$ are well-defined (by the relation (2.1)) for functions f which are integrable on T. (The term "integrable" may be interpreted as Riemann integrable or Lebesgue integrable depending on your knowledge). Prove the Riemann–Lebesgue Lemma (and its corollaries) for integrable functions f. Hint: If f is integrable, then for every $\varepsilon > 0$ there exists a continuous function g such that

$$\int_{-\pi}^{\pi} |f(t) - g(t)|\, dt < \varepsilon.$$

Notice, in particular, that piecewise C^1 functions are integrable and this will give you another proof of the second part of Exercise 2.3.4.

These corollaries too are sometimes called the Riemann–Lebesgue Lemma.

Theorem 2.3.6. *Let f be a continuous (or more generally, an integrable) function on T. Let $0 < \delta < \pi$. Then for every θ,*

$$\lim_{n \to \infty} \left(\int_{-\pi}^{-\delta} + \int_{\delta}^{\pi} \right) f(\theta - t) D_n(t)\, dt = 0. \qquad (2.12)$$

Proof Fix θ, and define a function g on T as

$$g(t) = \begin{cases} 0 & \text{if } |t| < \delta, \\ \dfrac{f(\theta - t)}{\sin t/2} & \text{if } \delta \leq |t| \leq \pi. \end{cases}$$

By Proposition 2.2.2(iii),

$$\left(\int_{-\pi}^{-\delta} + \int_{\delta}^{\pi} \right) f(\theta - t) D_n(t)\, dt = \frac{1}{2\pi} \int_{-\pi}^{\pi} g(t)\, (\sin(n + 1/2)t)\, dt.$$

Now note that g is integrable and use Exercise 2.3.3 and its extension in Exercise 2.3.5. ∎

Remark 2.3.7. In particular, choosing f to be the constant function 1 we get

$$\lim_{n \to \infty} \left(\int_{-\pi}^{-\delta} + \int_{\delta}^{\pi} \right) D_n(t)\, dt = 0.$$

This property is like Property (iv) of a Dirac sequence Q_n defined in Chapter 1. However D_n is *not* a Dirac sequence. What this property signifies is that as n becomes large the oscillations of D_n outside any δ-neighbourhood of 0 cancel each other out.

Remark 2.3.8. Theorem 2.3.6 is called the *Principle of Localisation.* Since

$$S_n(f; \theta) = \int_{-\pi}^{\pi} f(\theta - t) D_n(t) \, dt,$$

(2.12) shows that the convergence properties of f at θ are completely determined by the values of f in a δ-neighbourhood of θ, where δ can be arbitrarily small. So the study of convergence of $S_n(f; \theta)$ is reduced to that of the integral

$$\int_{-\delta}^{\delta} f(\theta - t) D_n(t) \, dt$$

for arbitrarily small δ.

This principle can also be stated as follows: if f is zero in a neighbourhood of θ, then $S_n(f; \theta)$ converges to zero. Another statement expressing this is: if f and g are equal in some neighbourhood of θ, then the Fourier series of f and g at θ are either both convergent to the same limit or are both divergent in the same way.

Now we can prove one of the several theorems that ensure convergence of the Fourier series of f when f satisfies some conditions stronger than continuity.

We say that f is *Lipschitz continuous* at θ if there exists a constant M and a $\delta > 0$ such that

$$|f(\theta) - f(t)| < M|\theta - t| \quad \text{if} \quad |\theta - t| < \delta. \tag{2.13}$$

This condition is stronger than continuity but weaker than differentiability of f at θ.

Theorem 2.3.9. *Let f be an integrable function on T. If f is Lipschitz continuous at θ, then*

$$\lim_{n \to \infty} S_n(f; \theta) = f(\theta).$$

Proof We want to prove

$$\lim_{n \to \infty} \int_{-\pi}^{\pi} f(\theta - t) D_n(t) \, dt = f(\theta).$$

By Proposition 2.2.2(ii) this amounts to showing

$$\lim_{n \to \infty} \int_{-\pi}^{\pi} [f(\theta - t) - f(\theta)] D_n(t) \, dt = 0. \tag{2.14}$$

Choose δ and M to satisfy (2.13). Using Proposition 2.2.2 (iii) observe that if $|t| < \delta$, then for all n,

$$\left| [f(\theta - t) - f(\theta)] D_n(t) \right| \le \frac{1}{\pi} M \left| \frac{t/2}{\sin t/2} \right|.$$

Hence for $0 < \varepsilon < \delta$ we have

$$\left| \int_{-\varepsilon}^{\varepsilon} [f(\theta - t) - f(\theta)] D_n(t)\, dt \right| \le C\varepsilon$$

for some constant C. Now, use Theorem 2.3.6 to get (2.14). ∎

Exercise 2.3.10. Let f be piecewise C^1 on T. Show that

$$\lim_{n \to \infty} S_n(f; \theta) = f(\theta) \quad \text{if } f \text{ is continuous at } \theta,$$

and

$$\lim_{n \to \infty} S_n(f; \theta) = \frac{f(\theta_+) + f(\theta_-)}{2}$$

if f is discontinuous at θ, and $f(\theta_+)$ and $f(\theta_-)$ are the right and the left limits of f at θ.
Hint: Write

$$S_n(f; \theta) = \int_0^{\pi} f(\theta + t) D_n(t)\, dt + \int_0^{\pi} f(\theta - t) D_n(t)\, dt,$$

$$S_n(f; \theta) - \frac{1}{2}[f(\theta_+) + f(\theta_-)] = \int_0^{\pi} [f(\theta + t) - f(\theta_+)] D_n(t)\, dt$$

$$+ \int_0^{\pi} [f(\theta - t) - f(\theta_-)] D_n(t)\, dt.$$

Then use the arguments of the proof of Theorem 2.3.9 to conclude that each of the above two integrals goes to zero as $n \to \infty$.

It will be worthwhile to understand how the additional condition on f in Theorem 2.3.9 has helped. In the proof of Theorem 1.3.1 and that of Theorem 2.3.9 we split the integral into two parts. However, in the first case since $Q_n(t) \ge 0$, taking absolute values did not affect it. In the case of $D_n(t)$, however, $\int |D_n(t)|\, dt$ becomes unbounded. In this case one of the integrals involved goes to zero because of the localisation principle which is a consequence of oscillations of $D_n(t)$. The other integral goes to zero because for small t the integrand in (2.14) is bounded *independently of n*. For this it is necessary that f itself should not be wildly oscillatory.

We will look at this from another angle also. As we have mentioned earlier, there exists a continuous function f for which the sequence

$$S_N(f; \theta) = \sum_{n=-N}^{N} \hat{f}(n) e^{in\theta}$$

diverges for some θ. In Chapter 1 we saw that for each real number $0 \le r < 1$ the series

$$\sum r^{|n|} \hat{f}(n) e^{in\theta}$$

converges. We can think of this as the insertion of a "convergence factor" $r^{|n|}$ which controls the size of the terms. It is true that $\hat{f}(n) \to 0$ by the Riemann–Lebesgue Lemma but not *fast enough* for the Fourier series to converge. The insertion of $r^{|n|}$ achieves this. What happens in Fejér's Theorem? Notice that we can write the "Fejér sum" $\sigma_N(f; \theta)$ as

$$\sigma_N(f; \theta) = \frac{1}{N} \sum_{k=0}^{N-1} S_k(f; \theta)$$

$$= \sum_{n=-(N-1)}^{N-1} \left(1 - \frac{|n|}{N}\right) \hat{f}(n) e^{in\theta}.$$

Theorem 2.2.5 says that $\sigma_N(f; \theta) \to f(\theta)$ as $N \to \infty$. So, here again the insertion of the "convergence factor" $1 - |n|/N$ seems to have helped. This suggests that if $\hat{f}(n)$ themselves go to zero fast enough, then the Fourier series of f might converge without any help from convergence factors.

Exercise 2.3.11. Let f be a C^1 function on T with derivative f'. Show that

$$\widehat{f'}(n) = in\hat{f}(n). \tag{2.15}$$

This is one of the most important facts of Fourier analysis whose importance you will discover as you proceed further.

Exercise 2.3.12. Let f be a C^1 function on T. Show that

$$\hat{f}(n) = O\left(\frac{1}{n}\right), \tag{2.16}$$

i.e., there exists a constant A such that

$$|\hat{f}(n)| \le \frac{A}{|n|}, n \ne 0. \tag{2.17}$$

Hint: Use (2.15) and the Riemann–Lebesgue Lemma. Prove (2.16) also when f is piecewise C^1.

Thus, whereas the Riemann–Lebesgue Lemma ensures only that $\hat{f}(n) \to 0$, under the additional hypothesis of f having a continuous derivative $\hat{f}(n) \to 0$ at least as fast as $1/n$. In fact the smoother f is the faster is the decay of $\hat{f}(n)$:

Exercise 2.3.13. Show that f is a function of class C^k on T (i.e., f has continuous derivatives up to order k), then $\hat{f}(n) = O(1/n^k)$.

Again, it should be emphasized that this relation between the smoothness of f and the size of its Fourier coefficients is an important fact of Fourier analysis.

Another important class of functions for which (2.16) holds is the functions of bounded variation. If P is a partition of $[a, b]$, i.e., a subdivision of this interval as

$$a = t_0 < t_1 < t_2 < \cdots < t_n = b,$$

let

$$v(f, P) = \sum_{i=1}^{n} |f(t_i) - f(t_{i-1})|. \tag{2.18}$$

If this sum is less than a fixed number K for every partition P, we say that f is of *bounded variation* on $[a, b]$ and then its *total variation* is defined as

$$V(f) = \sup v(f, P), \tag{2.19}$$

where the supremum is taken over all partitions P.

Exercise 2.3.14.

(i) A function f on an interval $[a, b]$ is said to be *uniformly Lipschitz* if there exists a constant K such that

$$|f(s) - f(t)| \le K|s - t| \quad \text{for all } s, t \text{ in } [a, b].$$

Show that every such function is of bounded variation on $[a, b]$.

(ii) Suppose f is a continuous function on $[a, b]$ and has a bounded derivative on (a, b). Then f is uniformly Lipschitz on $[a, b]$.

(iii) Any piecewise C^1 function on $[a, b]$ is of bounded variation.

(iv) The function

$$f(t) = \begin{cases} 0, & t = 0, \\ t \cos(\pi/t), & 0 < t \le 1 \end{cases}$$

is continuous (and hence uniformly continuous) on $[0, 1]$, but not of bounded variation. (Consider the partitions

$$0 < \frac{1}{n} < \frac{1}{n-1} < \cdots < \frac{1}{2} < 1.)$$

(v) If f is of bounded variation on $[a, b]$, then there exist monotonically increasing functions g, h such that $f = g - h$.

Lemma 2.3.15. *If f is a continuous function of bounded variation on T, then*

$$\hat{f}(n) = O\left(\frac{1}{n}\right).$$

Proof Integrate by parts using Riemann–Stieltjes integrals:

$$|\hat{f}(n)| = \left| \frac{1}{2\pi} \int_{-\pi}^{\pi} f(t) e^{-int} \, dt \right| = \left| \frac{1}{2\pi in} \int_{-\pi}^{\pi} e^{-int} \, df(t) \right| \le \frac{V(f)}{2\pi n}. \qquad \blacksquare$$

Exercise. Prove this when f is piecewise continuous.

For us this information is useful when we apply the following important result:

Theorem 2.3.16. *Let f be an integrable function on T such that $\hat{f}(n) = O(1/n)$. Then $S_n(f; \theta) \to f(\theta)$ at all points θ at which f is continuous. If f is continuous on T, this convergence is uniform.*

To prove this we will use a sum which is in between the two sums $S_n(f; \theta)$ and $\sigma_n(f; \theta)$ introduced earlier. For each pair of integers m, n, where $0 \le m < n$, define

$$\sigma_{m,n}(f; \theta) = \frac{S_{m+1}(f; \theta) + \cdots + S_n(f; \theta)}{n - m}. \tag{2.20}$$

Note that we can write

$$\sigma_{m,n}(f; \theta) = \frac{(n + 1)\sigma_{n+1}(f; \theta) - (m + 1)\sigma_{m+1}(f; \theta)}{n - m}, \tag{2.21}$$

and also

$$\sigma_{m,n}(f; \theta) = S_m(f; \theta) + \sum_{m < |j| \le n} \frac{n + 1 - |j|}{n - m} \hat{f}(j)e^{ij\theta}. \tag{2.22}$$

One way to see that this is "in between" S_n and σ_n is to write each of these sums as $\sum_{j=-n}^{n} \alpha(j)\hat{f}(j)e^{ij\theta}$. For the sum $S_n(f; \theta)$ the coefficients $\alpha(j)$ are 1 for $-n \le j \le n$ and 0 for $|j| > n$. For the sum $\sigma_n(f; \theta)$ the coefficients $\alpha(j)$ drop from the value 1 at $j = 0$ to the value 0 at $|j| = n$ and stay 0 after that. For $\sigma_{m,n}(f; \theta)$ the coefficients $\alpha(j)$ are 1 for $|j| \le m$ and then drop to 0 at $|j| = n + 1$ and stay 0 after that. (If you sketch the graph of $\alpha(j)$ you will get a rectangle, a triangle and a trapezium, respectively, in the three cases.)

We will look at the special case $\sigma_{kn,(k+1)n}$ of these sums.

Lemma 2.3.17. *Let f be integrable on T. Then for each fixed integer k,*

$$\sigma_{kn,(k+1)n}(f; \theta) \to f(\theta) \quad as \ n \to \infty$$

at each θ where f is continuous. If f is continuous on T this convergence is uniform on T.

Proof From the relation (2.21) write

$$\sigma_{kn,(k+1)n}(f; \theta) = (k + 1 + 1/n)\sigma_{(k+1)n}(f; \theta) - (k + 1/n)\sigma_{kn}(f; \theta).$$

Then use Fejér's Theorem (Theorem 2.2.5). Note that you will need to generalise this theorem to be able to prove the first statement of this lemma. This is left as an exercise. ∎

Lemma 2.3.18. *Let f be integrable on T and let $|\hat{f}(j)| \le A/|j|$ for $j \ne 0$. Then for all positive integers k, m, n with $kn \le m < (k + 1)n$ we have*

$$|\sigma_{kn,(k+1)n}(f; \theta) - S_m(f; \theta)| \le \frac{2A}{k}$$

Proof From the relation (2.22) we get

$$|\sigma_{kn,(k+1)n}(f;\theta) - S_m(f;\theta)| \leq \sum_{kn<|j|\leq(k+1)n} |\hat{f}(j)|$$

$$\leq 2 \sum_{j=kn+1}^{(k+1)n} \frac{A}{j}$$

$$\leq \frac{2nA}{kn} \leq \frac{2A}{k}. \qquad \blacksquare$$

Proof of Theorem 2.3.16 We are given that there exists a constant A such that $|\hat{f}(j)| \leq A/|j|$ for $j \neq 0$. Let $\varepsilon > 0$ be given. Choose an integer k such that $A/k < \varepsilon/4$. By Lemma 2.3.17 we can find $n_0 \geq k$ such that for all $n \geq n_0$,

$$|\sigma_{kn,(k+1)n}(f;\theta) - f(\theta)| < \frac{\varepsilon}{2}.$$

Now let $m \geq kn_0$. Then for some $n \geq n_0$ we will have $kn \leq m < (k+1)n$. Hence by Lemma 2.3.18,

$$|\sigma_{kn,(k+1)n}(f;\theta) - S_m(f;\theta)| \leq \frac{2A}{k} < \frac{\varepsilon}{2}.$$

From the above two inequalities it follows that

$$|S_m(f;\theta) - f(\theta)| < \varepsilon. \qquad \blacksquare$$

The most important corollaries of Theorem 2.3.16 are the following theorems:

Theorem 2.3.19. (Dirichlet's Theorem) *Let f be a piecewise C^1 function on T. Then the Fourier series $\sum \hat{f}(n)e^{in\theta}$ converges to $f(\theta)$ at every point θ where f is continuous. This convergence is uniform on any closed interval that does not contain a discontinuity of f.*

Theorem 2.3.20. (Jordan's Theorem) *The conclusion of Dirichlet's Theorem is valid, more generally, if f is any function of bounded variation on T.*

Exercise 2.3.21. Recall that a function of bounded variation is continuous except possibly at a countable set of points; and at these discontinuities the left and the right limits exist. Show that if θ is a point of discontinuity of f, then the Fourier series $\sum \hat{f}(n)e^{in\theta}$ converges to $\frac{1}{2}\{f(\theta_+) + f(\theta_-)\}$.

Let us make a few more remarks about the condition $\hat{f}(n) = O(1/n)$ under which we have proved our Theorem 2.3.16. In Chapter 3 we will see examples of functions that are piecewise C^1 (and hence, of bounded variation) for which the Fourier coefficients decay exactly as $1/n$. If f is a C^1 function on T, from the Riemann–Lebesgue Lemma and (2.15) we get a stronger result than (2.16): we have

$$\hat{f}(n) = o\left(\frac{1}{n}\right), \tag{2.23}$$

i.e., $\lim_{n\to\infty} \hat{f}(n)/n = 0$. There is another class of functions in between C^1 functions and functions of bounded variation that is useful, especially in the theory of Lebesgue integration. This is the class of absolutely continuous functions.

A function f on $[a, b]$ is said to be *absolutely continuous* if for every $\varepsilon > 0$, there exists $\delta > 0$ such that whenever $\{(a_i, b_i)\}$ is a finite disjoint collection of open intervals in $[a, b]$ with $\sum_{i=1}^{n}(b_i - a_i) < \delta$, then we have $\sum_{i=1}^{n}|f(b_i) - f(a_i)| < \varepsilon$.

Exercise 2.3.22. Prove the following statements.

(i) Every absolutely continuous function is uniformly continuous.

(ii) Every absolutely continuous function is of bounded variation.

(iii) If f is uniformly Lipschitz, then f is absolutely continuous.

(iv) In particular if f has a bounded derivative on (a, b), then f is absolutely continuous.

The *Fundamental Theorem of Calculus* says that if f is absolutely continuous on $[a, b]$, then it is differentiable almost everywhere, its derivative f' is integrable, and

$$f(t) = \int_a^t f'(s)\, ds + f(a) \quad \text{for all } a \le t \le b.$$

Conversely, if g is in $L_1[a, b]$, then the function $G(t) = \int_a^t g(s)\, ds$ is absolutely continuous, and $G' = g$ almost everywhere. (See e.g., H.L. Royden, *Real Analysis*.)

Exercise 2.3.23. Show that if f is absolutely continuous, then $\hat{f}(n) = o(1/n)$. In particular, this is true if f is a continuous function that is differentiable except at a finite number of points, and the derivative f' is bounded. (This is the case that arises most often in practice.)

Exercise 2.3.24. In Exercise 1.3.10 we saw that the convolution $f * g$ inherits the better of the smoothness properties of f and g. This idea goes further. Show that if g is of bounded variation, or is absolutely continuous, then $f * g$ has the same property.

We add, as a matter of record, that there exist continuous functions of bounded variation for which $\hat{f} \ne o(1/n)$. (The standard example of a continuous, but not absolutely continuous, function of bounded variation involves the Cantor set.)

2.4 Term by term integration and differentiation

If $\sum f_n(x) = f(x)$, and the series converges uniformly, we can obtain $\int f$ by integrating the series term by term. For Fourier series we can perform term by term integration even if the series does not converge at all.

Let f be any piecewise continuous (or, more generally, integrable) function on T. Let $c_n = \hat{f}(n)$. Let

$$F(x) = \int_0^x f(t)\, dt - c_0 x, \qquad -\pi \le x \le \pi.$$

Then F is absolutely continuous, $F(0) = 0$ and $F(\pi) = F(-\pi)$. Since F is absolutely continuous, its Fourier series

$$F(x) = \sum_{n=-\infty}^{\infty} A_n e^{inx}$$

converges uniformly. The coefficients c_n and A_n are related by the equation $c_n = in A_n$. From this we get

$$A_n = -i \frac{c_n}{n} \quad \text{for } n \ne 0.$$

The condition $F(0) = 0$ then shows

$$A_0 = i \sum_{n \ne 0} \frac{c_n}{n}.$$

This shows

$$\int_0^x f(t)\, dt = c_0 x + i \sum_{n \ne 0} \frac{c_n}{n} - i \sum_{n \ne 0} \frac{c_n}{n} e^{inx}. \tag{2.24}$$

This is exactly what we would have obtained on integrating the series

$$f(t) = \sum_{n=-\infty}^{\infty} c_n e^{int} \tag{2.25}$$

term by term from 0 to x. We have proved the following.

Theorem 2.4.1. *Let f be any function in $L^1(T)$ with Fourier series (2.25). Then the function $\int_0^x f(t)\, dt$ is represented by the series (2.24). The latter series is uniformly convergent on T, even though the former may be divergent at some points.*

Note that the series (2.24) is not quite a Fourier series because of the presence of the first term. Subtracting this term we get the Fourier series for the periodic function $F(x)$.

How about term by term differentiation?

Exercise 2.4.2. Let f be a continuous and piecewise C^1 function on T. Show that if f' is piecewise C^1, then the Fourier series $f(t) = \sum \hat{f}(n) e^{int}$ can be differentiated term by term, and the series so obtained converges pointwise to $\frac{1}{2}[f'(t_+) + f'(t_-)]$. (Use Dirichlet's Theorem.)

The conditions of Exercise 2.4.2 are satisfied by functions like $f(t) = |t|$ and $f(t) = |\sin t|$.

2.5 Divergence of Fourier series

We have informed the reader earlier that there exists a continuous function whose Fourier series diverges at some point. An example of such a function was constructed by Du Bois-Reymond in 1876. This example came as a surprise because it was generally believed by Dirichlet and, following him, by other mathematicians like Riemann and Weierstrass that the Fourier series of a continuous function should converge at every point.

To construct this example we make use of the fact that the Lebesgue constants L_n tend to ∞. We start with trigonometric polynomials with large Fourier sums, and combine them to get successively nastier functions whose limit behaves more wildly than any function whose formula could be explicitly written down. This idea is called the method of *condensation of singularities*.

Exercise 2.5.1.

(i) Let A be any positive real number. Choose N so that the Lebesgue constant $L_N > A$. Let $g_N(t) = \text{sgn} \, D_N(t)$; i.e., $g_N(t) = 1$ if $D_N(t) \geq 0$ and $g_N(t) = -1$ if $D_N(t) < 0$. Then the Fourier sum

$$S_N(g_N; 0) = L_N > A.$$

(ii) The function g_N is a step function having a finite number of discontinuities in $[-\pi, \pi]$. For each $\varepsilon > 0$ we can find a continuous function g such that $|g(t)| \leq 1$ and $\int_{-\pi}^{\pi} |g(t) - g_N(t)| \, dt < \varepsilon$. (Sketch a piecewise linear function with this property.) Use this to show that there exists a continuous function g on $[-\pi, \pi]$ with $|g(t)| \leq 1$ and $|S_N(g; 0)| > A$.

(iii) Use the Weierstrass approximation theorem to show that there exists a trigonometric polynomial p such that $|p(t)| \leq 1$ for all t in T and $|S_N(p; 0)| > A$. In other words, for each $A > 0$, there is a natural number N and a trigonometric polynomial $p(t) = \sum_{n=-M}^{M} \hat{p}(n)e^{int}$, such that

$$\left| \sum_{n=-M}^{M} \hat{p}(n)e^{int} \right| \leq 1 \quad \text{for all } t \in T,$$

but

$$\left| \sum_{n=-N}^{N} \hat{p}(n) \right| > A.$$

Theorem 2.5.2. (Du Bois-Reymond) *There exists a continuous function f on T whose Fourier series diverges at the point 0.*

Proof Using Exercise 2.5.1 (iii) we can find, for each $k = 1, 2, \ldots$, a trigonometric polynomial

$$p_k = \sum_{j=-m(k)}^{m(k)} \hat{p}_k(j) e^{ijt} \tag{2.26}$$

such that

$$|p_k(t)| \le 1 \quad \text{for all } t \in T \tag{2.27}$$

and a positive integer $n(k)$ such that

$$\left| \sum_{j=-n(k)}^{n(k)} \hat{p}_k(j) \right| \ge 2^{2k}. \tag{2.28}$$

We may assume, by adding zero terms if necessary, that $m(k) > n(k)$ and $m(k) \ge m(k-1)$. Now let

$$r(k) = \sum_{j=1}^{k} [2m(j) + 1]$$

and

$$f_n(t) = \sum_{k=1}^{n} \frac{1}{2^k} e^{ir(k)t} p_k(t)$$

$$= \sum_{k=1}^{n} \sum_{j=-m(k)}^{m(k)} \frac{1}{2^k} e^{i[r(k)+j]t} \hat{p}_k(j). \tag{2.29}$$

From this it is clear that if $n \ge k$ and $|j| \le m(k)$, then

$$\hat{f}_n(r(k) + j) = \frac{1}{2^k} \hat{p}_k(j), \tag{2.30}$$

and

$$\hat{f}_n(j) = 0 \quad \text{for all } j < 0. \tag{2.31}$$

If $n' \ge n + 1$, then

$$|f_{n'}(t) - f_n(t)| = \left| \sum_{k=n+1}^{n'} \frac{1}{2^k} e^{ir(k)t} p_k(t) \right|$$

$$\le \sum_{k=n+1}^{n'} \frac{1}{2^k} |p_k(t)|$$

$$= \sum_{k=n+1}^{n'} \frac{1}{2^k}.$$

So by the Weierstrass M-test the sequence f_n converges uniformly on T to a continuous function f. The properties (2.30) and (2.31) are carried over to f. Using these two properties we see that

$$\left| S_{r(k)+n(k)}(f; 0) - S_{r(k)-n(k)}(f; 0) \right|$$

$$= \left| \sum_{j=0}^{r(k)+n(k)} \hat{f}(j) - \sum_{j=0}^{r(k)-n(k)} \hat{f}(j) \right|$$

$$= \left| \sum_{j=-n(k)}^{n(k)} \hat{f}(r(k) + j) - \hat{f}(r(k) - n(k)) \right|$$

$$= \frac{1}{2^k} \left| \sum_{j=-n(k)}^{n(k)} \hat{p}_k(j) - \hat{p}_k(n(k)) \right|$$

$$\geq \frac{1}{2^k} \left| \sum_{j=-n(k)}^{n(k)} \hat{p}_k(j) \right| - \left| \hat{p}_k(n(k)) \right|$$

$$\geq \frac{1}{2^k}(2^{2k} - 1)$$

using (2.27) and (2.28). As $k \to \infty$ this expression goes to ∞. Thus the sequence $S_N(f; 0)$ cannot converge. ∎

Note that our proof shows that $\overline{\lim} \, S_N(f; 0) = \infty$.

The method of condensation of singularities has been found to be very useful in other contexts. Some of its essence is captured in one of the basic theorems of functional analysis called the Uniform Boundedness Principle or the Banach–Steinhaus Theorem (discovered by Lebesgue in 1908 in connection with his work on Fourier series; and made into a general abstract result by Banach and Steinhaus). We state the principle and then show how it can be used to give another proof of Theorem 2.5.2.

The Uniform Boundedness Principle. *Let X and Y be Banach spaces and let Λ_n be a sequence of bounded linear operators from X to Y. If the sequence $\|\Lambda_n x\|$ is bounded for each x in X, then the sequence $\|\Lambda_n\|$ is also bounded.*

Let X be the Banach space $C(T)$ consisting of continuous functions on T with the norm of such a function defined as

$$\|f\| = \sup_{t \in T} |f(t)|. \tag{2.32}$$

Now define a sequence of linear functionals on X as

$$\Lambda_n(f) = S_n(f; 0). \tag{2.33}$$

Note that

$$|\Lambda_n(f)| = \left| \int_{-\pi}^{\pi} f(t) D_n(t) \, dt \right|$$

$$\leq \|f\| \int_{-\pi}^{\pi} |D_n(t)| \, dt$$

$$= L_n \|f\|, \tag{2.34}$$

where L_n is the Lebesgue constant. In particular this shows that

$$\|\Lambda_n\| \leq L_n \quad \text{for each } n. \tag{2.35}$$

We will show that

$$\|\Lambda_n\| = L_n \quad \text{for each } n. \tag{2.36}$$

Fix n. Let $g(t) = \operatorname{sgn} D_n(t)$. As seen in Exercise 2.5.1, we can choose a sequence ϕ_m in $C(T)$ such that $|\phi_m(t)| \leq 1$ and $\lim_{m \to \infty} \phi_m(t) = g(t)$ for every t. Hence, by the Dominated Convergence Theorem,

$$\lim_{m \to \infty} \Lambda_n(\phi_m) = \lim_{m \to \infty} \int_{-\pi}^{\pi} \phi_m(t) D_n(t) \, dt$$

$$= \int_{-\pi}^{\pi} g(t) D_n(t) \, dt$$

$$= \int_{-\pi}^{\pi} |D_n(t)| \, dt.$$

Since $\|\phi_m\| = 1$, this proves (2.36). So $\|\Lambda_n\|$ is not a bounded sequence. Hence by the Uniform Boundedness Principle there exists an f in the space $X = C(T)$ such that the sequence

$$|\Lambda_n(f)| = |S_n(f; 0)|$$

is not bounded. Therefore, the Fourier series of f at 0 does not converge.

Let us make a few remarks here. The point 0 was chosen just for convenience. The same argument shows that for each point θ on T we can find a function f such that $|S_n(f; \theta)|$ is not bounded. The Uniform Boundedness Principle is (usually) proved using the Baire Category Theorem. Using these methods one can see that the set of all functions whose Fourier series converge at 0 is a set of first category in the space $C(T)$. In this sense continuous functions whose Fourier series converge everywhere form a *meagre set*. Using a little more delicate analysis one can show the existence of a continuous function whose Fourier series diverges except on a set of points of the first category in T, and the existence of a continuous function whose Fourier series diverges on an uncountable set. (See W. Rudin, *Real and Complex Analysis,* Chapter 5).

One may now wonder whether it is possible to have a continuous function whose Fourier series diverges *everywhere*.

After Du Bois-Reymond's result, this is what mathematicians tried to prove for several years. In 1926, A.N. Kolmogorov constructed a Lebesgue integrable function on T, i.e., a function in the class $L^1(T)$, such that $|S_n(f; \theta)|$ diverges for all θ. Though Kolmogorov's function was not Riemann integrable, this encouraged people in their search for a continuous function whose Fourier series might diverge everywhere. However, in 1966, L. Carleson proved that if f is square integrable, i.e., in the space $L^2(T)$, then the Fourier series of f converges to f almost everywhere on T. In particular if f is continuous, then its Fourier series converges almost everywhere.

Another natural question may be raised at this stage. Given a set E of measure zero in T, can one find a continuous function f on T such that $|S_n(f; \theta)|$ is divergent for all θ in E? The answer is yes! This was shown by J.P. Kahane and Y. Katznelson soon after Carleson's result was proved.

So Fourier was wrong in believing that his series for a continuous function will converge at every point. But he has been proved to be *almost* right.

3

Odds and Ends

3.1 Sine and cosine series

A function f is called *odd* if $f(x) = -f(-x)$ and *even* if $f(x) = f(-x)$ for all x. An arbitrary function f can be decomposed as $f = f_{\text{even}} + f_{\text{odd}}$, where $f_{\text{even}}(x) = \frac{1}{2}[f(x) + f(-x)]$ and $f_{\text{odd}}(x) = \frac{1}{2}[f(x) - f(-x)]$. The functions f_{even} and f_{odd} are called the *even part* and the *odd part* of f, respectively. Notice that if $f(x) = e^{ix}$, then $f_{\text{even}}(x) = \cos x$ and $f_{\text{odd}}(x) = i \sin x$.

Exercise 3.1.1. Show that if f is an even function on T, then $\hat{f}(n) = \hat{f}(-n)$, and if f is odd, then $\hat{f}(n) = -\hat{f}(-n)$. In other words the Fourier coefficients are also then even and odd functions on the integers Z.

So if f is an even function on T, then

$$S_N(f; \theta) = \sum_{n=-N}^{N} \hat{f}(n) e^{in\theta}$$

$$= \hat{f}(0) + 2 \sum_{n=1}^{N} \hat{f}(n) \cos n\theta.$$

Also note that if f is even, then

$$\hat{f}(n) = \frac{1}{2\pi} \int_{-\pi}^{\pi} f(t)(\cos nt - i \sin nt) \, dt$$

$$= \frac{1}{\pi} \int_{0}^{\pi} f(t) \cos nt \, dt.$$

Putting $a_n = 2\hat{f}(n)$, we can write the Fourier series of an even function f as

$$\frac{a_0}{2} + \sum_{n=1}^{\infty} a_n \cos n\theta, \tag{3.1}$$

where

$$a_n = \frac{1}{\pi} \int_{-\pi}^{\pi} f(t) \cos nt \, dt. \tag{3.2}$$

This is called the *Fourier cosine series*.

In the same way, if f is an odd function on T, then we can write its *Fourier sine series* as

$$\sum_{n=1}^{\infty} b_n \sin n\theta \tag{3.3}$$

where,

$$b_n = \frac{1}{\pi} \int_{-\pi}^{\pi} f(t) \sin nt \, dt. \tag{3.4}$$

The Fourier series of any function on T can be expressed as

$$\frac{a_0}{2} + \sum_{n=1}^{\infty} (a_n \cos n\theta + b_n \sin n\theta), \tag{3.5}$$

where a_n and b_n are as given above. In several books the series (3.5) is called the Fourier series of f and is taken as the starting point. Let us say that (3.5) is the Fourier series in its *trigonometric form* and

$$\sum_{n=-\infty}^{\infty} c_n e^{in\theta} \tag{3.6}$$

is the Fourier series in its *exponential form*.

Exercise 3.1.2. If (3.5) and (3.6) are the Fourier series of the function f, show that

$$a_n = c_n + c_{-n}, \qquad b_n = i(c_n - c_{-n}),$$

and for $n \geq 0$,

$$c_n = \frac{1}{2}(a_n - ib_n), \qquad c_{-n} = \frac{1}{2}(a_n + ib_n).$$

Exercise 3.1.3. The series (3.5) can be integrated term by term to give

$$\int_0^x f(\theta) \, d\theta = \frac{a_0 x}{2} + \sum_{n=1}^{\infty} \frac{b_n}{n} + \sum_{n=1}^{\infty} \left(\frac{a_n}{n} \sin nx - \frac{b_n}{n} \cos nx \right). \tag{3.7}$$

If f is sufficiently smooth, the series (3.5) can be differentiated term by term, and then

$$f'(\theta) = \sum_{n=1}^{\infty} (nb_n \cos n\theta - na_n \sin n\theta). \tag{3.8}$$

(See Section 2.4)

Exercise 3.1.4. We have seen above that if f is a continuous even function on T, then given an $\varepsilon > 0$ there exists a finite sum $g(\theta) = \sum_{n=0}^{N} a_n \cos n\theta$ such that

$$\sup_{\theta \in T} |f(\theta) - g(\theta)| < \varepsilon. \tag{3.9}$$

This can be used to prove the *Weierstrass approximation theorem* as follows:

(i) Show, by induction, that there exists a polynomial T_n of degree n with real coefficients such that

$$\cos n\theta = T_n(\cos \theta). \tag{3.10}$$

T_n is called the Tchebychev polynomial of degree n.

(ii) Let φ be any continuous function on $[0, 1]$. Define a function f on T by $f(\theta) = \varphi(|\cos \theta|)$. Then f is an even continuous function. Use (3.9) and (3.10) to show that

$$\sup_{0 \leq t \leq 1} \left| \varphi(t) - \sum_{n=0}^{N} a_n T_n(t) \right| < \varepsilon.$$

This shows that every continuous function on $[0, 1]$ is a uniform limit of polynomials.

(iii) By a change of variables show that every continuous function on an interval $[a, b]$ is a uniform limit of polynomials. This is called the Weierstrass approximation theorem.

(iv) Show that this result and the one we proved in Exercise 1.3.9 can be derived from each other.

Exercise 3.1.5. Use the Weierstrass approximation theorem to prove that if f and g are continuous functions on $[a, b]$ such that

$$\int_a^b x^n f(x) \, dx = \int_a^b x^n g(x) \, dx$$

for all $n \geq 0$, then $f = g$. (This is called the *Hausdorff moment theorem* and is very useful in probability theory).

3.2 Functions with arbitrary periods

So far, we have considered periodic functions with period 2π, identified them with functions on T and obtained their Fourier series. A very minor modification is required to deal with functions of period $2L$, where L is any real number. If f is periodic with period $2L$, then the function $g(x) = f(xL/\pi)$ is periodic, with period 2π. So, we can first derive the Fourier series for g and then translate everything back to f by a change of variables. All the results proved earlier remain valid with appropriate modifications.

Exercise 3.2.1. Let L be any positive number and let f be an integrable function on $[-L, L]$ with $f(L) = f(-L)$. Then the Fourier series of f can be written as

$$\sum \hat{A}_n e^{inx\pi/L},$$

where

$$\hat{A}_n = \frac{1}{2L} \int_{-L}^{L} f(t) e^{-int\pi/L} \, dt. \tag{3.11}$$

In the same way, if f is even, then its cosine series can be written as

$$\frac{1}{2} A_0 + \sum_{n=1}^{\infty} A_n \cos \frac{nx\pi}{L},$$

where

$$A_n = \frac{2}{L} \int_0^L f(t) \cos \frac{nt\pi}{L} \, dt; \tag{3.12}$$

and if f is odd, then its sine series can be written as

$$\sum_{n=1}^{\infty} B_n \sin \frac{nx\pi}{L},$$

where

$$B_n = \frac{2}{L} \int_0^L f(t) \sin \frac{nt\pi}{L} \, dt. \tag{3.13}$$

Now we come to an interesting bit: how to apply these ideas to *any* function on a bounded interval. Let f be any continuous (or piecewise C^1) function defined on the interval $(0, L)$. Extend f to $[-L, L]$ as follows. Define

$$f(0) = f(0_+),$$

$$f(-x) = f(x) \quad \text{for } 0 < x \le L.$$

This gives a continuous (or piecewise C^1) function on $[-L, L]$. Now extend f to all of $\mathbb{R}$ by putting

$$f(x + 2L) = f(x) \quad \text{for all } x.$$

This is an even continuous (or piecewise C^1) function defined on all of $\mathbb{R}$ having period $2L$. So we can expand f into a cosine series. Notice that our extended function is continuous at $x = 0$ and $x = L$. We could also have extended f to an odd function as follows. Define

$$f(0) = 0,$$

$$f(-x) = -f(x), \qquad 0 < x \le L.$$

This extends f to $[-L, L]$; and now extend f to $\mathbb{R}$ by putting

$$f(x + 2L) = f(x) \quad \text{for all } x.$$

Now we get an odd periodic function. Notice that now $f(0) = 0$ and f is discontinuous at L with $f(L_-) = -f(L_+)$. So we can expand f into a sine series also.

   This apparently anomalous situation—the possibility of expanding any function as a sum of periodic functions and further the possibility of expanding the same function as both a sum of even functions and a sum of odd functions—was a part of the raging controversy at the beginning of the subject. See Chapter 0.

   Examples in the next section will clarify further how this apparent anomaly is neatly resolved by our analysis.

3.3 Some simple examples

Example 3.3.1. Let f be the "*pulse function*" defined as

$$f(x) = \begin{cases} -1 & \text{if } -\pi \le x < 0, \\ 1 & \text{if } 0 \le x < \pi, \end{cases}$$

and extended periodically to all of $\mathbb{R}$. This is an odd piecewise C^1 function with jumps at $0, \pm\pi, \pm 2\pi, \ldots$. The average value of f at these points is 0. The Fourier series of f is a sine series like (3.3) with coefficients given by

$$b_n = \frac{2}{\pi} \int_0^\pi \sin nt \, dt.$$

So

$$b_n = \begin{cases} 4/n\pi & \text{if } n \text{ is odd}, \\ 0 & \text{if } n \text{ is even}. \end{cases}$$

Hence we have

$$f(x) = \frac{4}{\pi} \left(\frac{\sin x}{1} + \frac{\sin 3x}{3} + \frac{\sin 5x}{5} + \cdots \right). \tag{3.14}$$

At $x = 0, \pm\pi, \pm 2\pi, \ldots$, the series adds up to zero which is the average value of f at these points. At other points it converges to $f(x)$. Figure 1 shows the pulse function. Its Fourier sums for 1, 2, 10 and 100 terms are shown in the next figure.

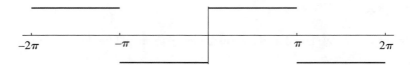

FIGURE 1
The pulse function

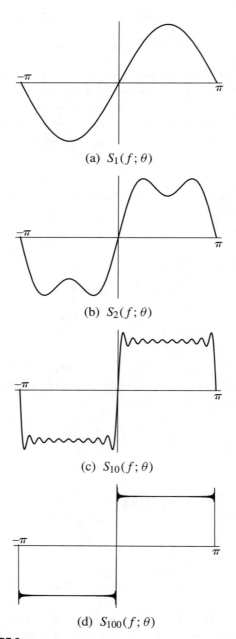

(a) $S_1(f; \theta)$

(b) $S_2(f; \theta)$

(c) $S_{10}(f; \theta)$

(d) $S_{100}(f; \theta)$

FIGURE 2
Fourier sums of the pulse function for different values of N

Example 3.3.2. Now consider the pulse function with a different width. Let

$$f(x) = \begin{cases} -1 & \text{for } -1 \le x < 0, \\ 1 & \text{for } 0 \le x < 1 \end{cases}$$

and extend it to $\mathbb{R}$ as a function with period 2. Show that the Fourier series of f is

$$f(x) = \frac{4}{\pi} \left(\frac{\sin \pi x}{1} + \frac{\sin 3\pi x}{3} + \frac{\sin 5\pi x}{5} + \cdots \right). \tag{3.15}$$

Note that the Fourier coefficients of f in Examples 3.3.1 and 3.3.2 are the same, but the wavelengths of the sine waves have been adjusted. (The *amplitude* of the wave is the same as before, the *wavelength* is shorter.)

Example 3.3.3. The function

$$f(x) = \begin{cases} 0 & \text{for } -1 \le x < 0, \\ 1 & \text{for } 0 \le x < 1 \end{cases}$$

is called the *Heaviside function.* Extend it to $\mathbb{R}$ as a function with period 2, and show its Fourier series is

$$\frac{1}{2} + \frac{2}{\pi} \left(\frac{\sin \pi x}{1} + \frac{\sin 3\pi x}{3} + \frac{\sin 5\pi x}{5} + \cdots \right). \tag{3.16}$$

Obtain this directly from the preceding example.

Example 3.3.4. Let

$$f(x) = \begin{cases} \dfrac{\pi}{2} + x & \text{for } -\pi \le x \le 0, \\ \dfrac{\pi}{2} - x & \text{for } 0 \le x \le \pi \end{cases}$$

and extend f to $\mathbb{R}$ as a periodic function with period 2π. You get a *"triangular wave"* with amplitude $\pi/2$ and period 2π. Show that f has the expansion

$$f(x) = \frac{4}{\pi} \left(\frac{\cos x}{1^2} + \frac{\cos 3x}{3^2} + \cdots \right). \tag{3.17}$$

Example 3.3.5. Let f be defined on $[0, 1)$ as

$$f(x) = x, \qquad 0 \le x < 1.$$

Extend f to $\mathbb{R}$ as an *even* function by the prescription in Section 3.2. This gives a *saw-tooth curve* shown in Figure 3. This is an even continuous function with the cosine series

$$f(x) = \frac{1}{2} - \frac{4}{\pi^2} \left(\frac{\cos \pi x}{1^2} + \frac{\cos 3\pi x}{3^2} + \cdots \right). \tag{3.18}$$

If we extend f to an odd function by our prescription we get the function shown in Figure 4.

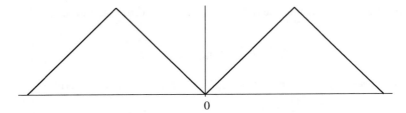

FIGURE 3
The even saw-tooth curve

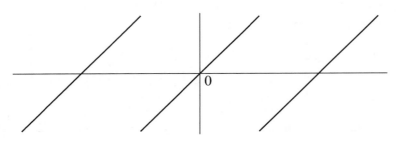

FIGURE 4
The odd saw-tooth curve

This has discontinuities at all odd integers, where the average value is zero. The Fourier series is now a sine series

$$f(x) = \frac{2}{\pi}\left(\frac{\sin \pi x}{1} - \frac{\sin 2\pi x}{2} + \frac{\sin 3\pi x}{3} - \cdots\right). \tag{3.19}$$

We can extend f beyond its original domain in one more way. We put $f(x) = 1 + x$ for $-1 \le x < 0$, and then extend the function to all of $\mathbb{R}$ as a periodic function with period 2. This gives the function (neither even nor odd), see Figure 5(a).

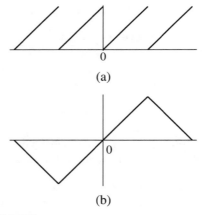

(a)

(b)

FIGURE 5
Two more extensions of the saw-tooth function

We can obtain the Fourier series for this function from (3.19) in three steps. First consider the odd function $g(x) = x$, $-\frac{1}{2} < x < \frac{1}{2}$, and extend it to $\mathbb{R}$ as a periodic function with period 1. The amplitude and the wavelength of g are half those of the function f in (3.19). So

$$g(x) = \frac{1}{\pi}\left(\frac{\sin 2\pi x}{1} - \frac{\sin 4\pi x}{2} + \frac{\sin 6\pi x}{3} - \cdots\right).$$

The function f whose Fourier series we want can be expressed as $f(x) = g(x - \frac{1}{2}) + \frac{1}{2}$. So its Fourier series is

$$f(x) = \frac{1}{2} - \frac{1}{\pi}\left(\frac{\sin 2\pi x}{1} + \frac{\sin 4\pi x}{2} - \frac{\sin 6\pi x}{3} - \cdots\right). \tag{3.20}$$

A fourth way of extending the original function of this example is as follows. Let

$$f(x) = \begin{cases} x & \text{for } -1 \le x \le 1, \\ 2 - x & \text{for } 1 \le x \le 2, \\ -2 - x & \text{for } -2 \le x \le -1. \end{cases}$$

Then extend f to all of $\mathbb{R}$ as a periodic function with period 4. This gives an odd triangular wave, see Figure 5(b). A calculation shows that f has the Fourier series

$$f(x) = \frac{8}{\pi^2}\left(\frac{\sin \pi x/2}{1^2} - \frac{\sin 3\pi x/2}{3^2} + \frac{\sin 5\pi x/2}{5^2} - \cdots\right). \tag{3.21}$$

We have thus four different periodic functions that coincide with $f(x)$, for $0 \le x < 1$. Their Fourier series are given by the formulas (3.18)–(3.21). The first function is even, the second and the fourth are odd, the third is neither even nor odd. Two of the four functions are continuous. The corresponding series (3.18) and (3.21) are uniformly convergent. The series (3.19) and (3.20) converge uniformly on closed subintervals of $(-1, 1)$ and $(0, 1)$ respectively. At the points $x = n$, where n is an integer, the series (3.19) converges to the value 0, and the series (3.20) to the value $1/2$.

Example 3.3.6. Let

$$f(x) = \begin{cases} -\cos x & \text{if } -\pi \le x < 0, \\ 0 & \text{if } x = 0, \\ \cos x & \text{if } 0 < x < \pi \end{cases}$$

and extend f to $\mathbb{R}$ as a periodic function with period 2π. This is an odd function with discontinuities at $x = n\pi$. Show that the Fourier series of f is given by

$$\frac{8}{\pi}\sum_{n=1}^{\infty}\frac{n\sin 2nx}{4n^2 - 1}. \tag{3.22}$$

If we differentiate this series term by term, we obtain

$$\frac{16}{\pi} \sum_{n=1}^{\infty} \frac{n^2 \cos 2nx}{4n^2 - 1}.$$

This series is divergent. (See Exercise 2.4.2).

Example 3.3.7. Let $f(x) = x^2$, $-\pi \le x \le \pi$, and extend this to $\mathbb{R}$ as a periodic function with period 2π. The Fourier cosine coefficients of this even function can be calculated by integrating by parts twice. We get

$$x^2 = \frac{\pi^2}{3} + 4 \sum_{n=1}^{\infty} (-1)^n \frac{\cos nx}{n^2} \qquad \text{for } -\pi \le x \le \pi. \tag{3.23}$$

This series can be differentiated term by term (see Exercise 2.4.2) to get

$$x = 2 \sum_{n=1}^{\infty} (-1)^{n+1} \frac{\sin nx}{n} \qquad \text{for } -\pi < x < \pi. \tag{3.24}$$

This is the series we have obtained in (3.19). In the reverse direction, we can integrate term by term the series obtained for x in Example 3.3.5 and obtain different series for x^2. Thus, the series (3.20) leads to

$$\frac{x^2}{2} = \pi x - 2 \sum_{n=1}^{\infty} \frac{1}{n^2} + 2 \sum_{n=1}^{\infty} \frac{\cos nx}{n^2} \qquad \text{for } 0 \le x \le 2\pi. \tag{3.25}$$

Integrating the series (3.18) we get

$$x^2 = \pi x - \frac{8}{\pi} \sum_{n=1}^{\infty} \frac{\sin nx}{n^3} \qquad \text{for } 0 \le x \le \pi. \tag{3.26}$$

These are not Fourier series. If we substitute in them the Fourier series for x (in the appropriate domains) given by (3.20) we obtain from (3.25)

$$x^2 = \frac{4\pi^2}{3} + 4 \sum_{n=1}^{\infty} \frac{\cos nx}{n^2} - 4\pi \sum_{n=1}^{\infty} \frac{\sin nx}{n}, \qquad \text{for } 0 < x < 2\pi, \tag{3.27}$$

using the fact $\sum 1/n^2 = \pi^2/6$ (see (3.39)), and from (3.26)

$$x^2 = \frac{\pi^2}{2} - \pi \sum_{n=1}^{\infty} \sin 2nx - \frac{8}{\pi} \sum_{n=1}^{\infty} \frac{\sin nx}{n^3}, \qquad \text{for } 0 < x < \pi. \tag{3.28}$$

Yet another series representation for x^2 is given in Exercise 3.5.2.

Example 3.3.8. Show that the Fourier series for the function $|\sin x|$ is

$$|\sin x| = \frac{2}{\pi} - \frac{4}{\pi} \sum_{n=1}^{\infty} \frac{\cos 2nx}{4n^2 - 1}. \tag{3.29}$$

This representation is valid for all $x \in \mathbb{R}$. Show that the Fourier series for the function

$$f(x) = \begin{cases} 0 & \text{for } -\pi \leq x \leq 0, \\ \sin x & \text{for } 0 \leq x \leq \pi \end{cases}$$

is

$$\frac{1}{\pi} + \frac{\sin x}{2} - \frac{2}{\pi} \sum_{n=1}^{\infty} \frac{\cos 2nx}{4n^2 - 1}. \tag{3.30}$$

3.4 Infinite products

Let $\{a_n\}$ be a sequence of real numbers. How do we define the infinite product Πa_n? We can imitate the definition of an infinite sum. For each $N = 1, 2, \ldots$ let $p_N = a_1 a_2 \cdots a_N$ be the sequence of *partial products* of $\{a_n\}$. If this sequence converges to a limit p and $p \neq 0$, we say that the product Πa_n converges, and write

$$\prod_{n=1}^{\infty} a_n = p.$$

Note that if any $a_n = 0$, then the product is zero. This case is uninteresting. Further, if p_N converges, then a_n too has to converge to 1. So, after ignoring the first few factors, we may assume that $a_n > 0$ for all n. In this case $\log a_n$ is meaningful, and we can convert questions about products to questions about sums.

Exercise 3.4.1. Let $a_n > 0$ for $n = 1, 2, \ldots$. Show that the product Πa_n converges if and only if the series $\sum \log a_n$ converges.

Exercise 3.4.2.

(i) Show that if $|x| < 1/2$, then

$$\frac{1}{2}|x| \leq |\log(1 + x)| \leq \frac{3}{2}|x|.$$

Hint: For $|x| < 1$, we have

$$\log(1 + x) = x - \frac{x^2}{2} + \frac{x^3}{3} - \frac{x^4}{4} + \cdots.$$

(ii) If $a_n > -1$ for all n, then the series $\sum |\log(1 + a_n)|$ converges if and only if $\sum |a_n|$ converges.

Let $\{a_n\}$ be a sequence of positive real numbers. If the series $\sum \log a_n$ converges absolutely, we say that the product Πa_n *converges absolutely*.

Exercise 3.4.3. Let $a_n > 0$ for all n. The product Πa_n converges absolutely if and only if the series $\sum (a_n - 1)$ converges absolutely.

Uniform convergence of products like $\Pi f_n(x)$ can be defined in the same way as for series.

We will derive now some important product representations for trigonometric functions. Consider the function

$$f(x) = \cos tx, \qquad -\pi \leq t \leq \pi, \qquad t \text{ not an integer.}$$

This is an even function. Its Fourier cosine series can be determined easily. We have

$$a_0 = \frac{2}{\pi} \frac{\sin t\pi}{t},$$

$$a_n = \frac{(-1)^n}{\pi} \frac{2t \sin t\pi}{t^2 - n^2}.$$

Thus we have

$$\cos tx = \frac{2t \sin t\pi}{\pi} \left(\frac{1}{2t^2} - \frac{\cos x}{t^2 - 1^2} + \frac{\cos 2x}{t^2 - 2^2} - \cdots \right).$$

Choosing $x = \pi$, we obtain

$$\frac{\cos t\pi}{\sin t\pi} = \frac{2t}{\pi} \left(\frac{1}{2t^2} + \frac{1}{t^2 - 1^2} + \frac{1}{t^2 - 2^2} + \cdots \right).$$

This can be rewritten as

$$\cot \pi t - \frac{1}{\pi t} = \frac{2t}{\pi} \sum_{n=1}^{\infty} \frac{1}{t^2 - n^2}, \qquad (3.31)$$

or as

$$\pi \cot \pi t = \frac{1}{t} + \sum_{n \neq 0} \left(\frac{1}{t - n} + \frac{1}{n} \right), \qquad (3.32)$$

where the last summation is over all nonzero integers.

This is a very important formula. It is called the *resolution of the cotangent into partial fractions.*

If $0 \leq t \leq a < 1$, the series in (3.31) is dominated by the convergent series

$$\sum \frac{1}{n^2 - a^2}.$$

Hence the series is uniformly convergent and can be integrated term by term in this domain. Performing this integration, we see that for $0 < x < 1$, we have

$$\log \frac{\sin \pi x}{\pi x} = \sum_{n=1}^{\infty} \log \left(1 - \frac{x^2}{n^2} \right).$$

Taking exponentials of both sides we get the marvelous product formula

$$\frac{\sin \pi x}{\pi x} = \prod_{n=1}^{\infty} \left(1 - \frac{x^2}{n^2}\right). \tag{3.33}$$

Exercise 3.4.4. Use the relation $\cos \pi x = \sin 2\pi x / (2 \sin \pi x)$ to obtain from this the product formula

$$\cos \pi x = \prod_{n=1}^{\infty} \left(1 - \frac{4x^2}{(2n-1)^2}\right). \tag{3.34}$$

Exercise 3.4.5. Show that

$$\frac{\pi^2}{\sin^2 \pi x} = \sum_{n=-\infty}^{\infty} \frac{1}{(x-n)^2}. \tag{3.35}$$

Exercise 3.4.6. We have obtained these infinite series and product expansions for $0 < x < 1$. Show that they are meaningful and true for all real x. (They are valid for all complex numbers, and are most often included in Complex Analysis courses.)

Exercise 3.4.7. Show that

$$\frac{1}{\sin x} = \frac{1}{x} + \sum_{n=1}^{\infty} (-1)^n \left[\frac{1}{x - n\pi} + \frac{1}{x + n\pi}\right].$$

3.5 π and infinite series

The number π has allured, fascinated and puzzled mathematicians for over two millenia. Approximations to π and formulas for it have been among the prize discoveries of famous mathematicians. Fourier series provide an inexhaustible source for such formulas. Some of them are given below. You can certainly discover more.

Choosing $x = \pi/2$ in (3.14) we get

$$\frac{\pi}{4} = 1 - \frac{1}{3} + \frac{1}{5} - \frac{1}{7} + \cdots. \tag{3.36}$$

This seems to be the earliest example of a series involving reciprocals of integers and π. It was discovered by Nilakantha in the fifteenth century via the series

$$\arctan x = x - \frac{x^3}{3} + \frac{x^5}{5} - \cdots, \tag{3.37}$$

valid for $|x| \le 1$. This series was known to the fourteenth century mathematician Madhava. These series are now generally known as Leibniz–Gregory series.

Choosing $x = 0$ in (3.17) we get

$$\frac{\pi^2}{8} = 1 + \frac{1}{3^2} + \frac{1}{5^2} + \cdots. \tag{3.38}$$

From this we can derive Euler's wonderful and famous formula discovered in 1734,

$$\frac{\pi^2}{6} = 1 + \frac{1}{2^2} + \frac{1}{3^2} + \cdots. \tag{3.39}$$

Note that the series on the right is convergent and its sum is bounded by 2. This was noticed by the Bernoulli brothers (Jakob and Johann) who asked for the exact value of the sum. Rearranging its terms we can write

$$\sum_{n=1}^{\infty} \frac{1}{n^2} = \sum_{n=1}^{\infty} \frac{1}{(2n-1)^2} + \sum_{n=1}^{\infty} \frac{1}{(2n)^2}$$

$$= \sum_{n=1}^{\infty} \frac{1}{(2n-1)^2} + \frac{1}{4} \sum_{n=1}^{\infty} \frac{1}{n^2}.$$

Using this we get (3.39) from (3.38). (Euler's proof was different and is discussed later.)
Choosing $x = 1/4$ we get from the series (3.19),

$$\frac{\pi}{2\sqrt{2}} = 1 + \frac{1}{3} - \frac{1}{5} - \frac{1}{7} + \frac{1}{9} + \frac{1}{11} - \cdots, \tag{3.40}$$

a series in which two negative signs alternate with two positive signs. Choosing $x = 1/2$ in (3.21) we get

$$\frac{\pi^2}{8\sqrt{2}} = 1 - \frac{1}{3^2} - \frac{1}{5^2} + \frac{1}{7^2} + \frac{1}{9^2} - \cdots. \tag{3.41}$$

Choosing $x = \pi/2$ in (3.26) we get

$$\frac{\pi^3}{32} = 1 - \frac{1}{3^3} + \frac{1}{5^3} - \frac{1}{7^3} + \cdots. \tag{3.42}$$

Choosing $x = \pi/2$ in (3.29) we get

$$\frac{\pi}{4} = \frac{1}{2} + \left(\frac{1}{1 \cdot 3} - \frac{1}{3 \cdot 5} + \frac{1}{5 \cdot 7} - \cdots \right). \tag{3.43}$$

Exercise 3.5.1. Show that

$$\sum_{n=1}^{\infty} \frac{1}{n^4} = \frac{\pi^4}{90}. \tag{3.44}$$

Exercise 3.5.2. Use (3.25), (3.39) and a rescaled version of (3.20) to show

$$x^2 = \frac{4}{3}\pi^2 + 4 \sum_{n=1}^{\infty} \left(\frac{\cos nx}{n^2} - \frac{\pi \sin nx}{n} \right), \tag{3.45}$$

for $0 < x < 2\pi$.

Exercise 3.5.3. From the product formula (3.33) obtain the *Wallis formula*

$$\frac{\pi}{2} = \frac{2}{1}\frac{2}{3}\frac{4}{3}\frac{4}{5}\cdots.$$

Exercise 3.5.4. Show that

$$\int_0^1 \frac{\log x}{1-x}\,dx = -\frac{\pi^2}{6}.$$

3.6 Bernoulli numbers

This is an important sequence of rational numbers discovered by Jakob Bernoulli. These numbers arise in several problems in analysis.

Let $B_n(x)$ be the sequence of functions on the interval $[0,1]$ defined inductively by the conditions

(i) $B_0(x) = 1$,

(ii) $B_n'(x) = B_{n-1}(x)$ for $n = 1, 2, \ldots$,

(iii) $\int_0^1 B_n(x)\,dx = 0$ for $n = 1, 2, \ldots$.

These conditions determine uniquely a sequence of polynomials called *Bernoulli polynomials*. The first four are given by

$$B_0(x) = 1,$$
$$B_1(x) = x - 1/2,$$
$$B_2(x) = x^2/2 - x/2 + 1/12,$$
$$B_3(x) = x^3/6 - x^2/4 + x/12,$$
$$B_4(x) = x^4/24 - x^3/12 + x^2/24 - 1/720.$$

Exercise 3.6.1. Show that

(i) $B_n(x)$ is a polynomial of degree n.

(ii) $B_n(x) = (-1)^n B_n(1-x)$ for all $n \geq 0$.

(iii) $B_n(0) = B_n(1)$ for all $n \geq 0$ except $n = 1$.

(Note $B_1(0) = -B_1(1) = -1/2$.)

The *Bernoulli numbers* are the sequence B_n defined as

$$B_n = n! B_n(0). \tag{3.46}$$

We should warn the reader here that different books use different conventions. Some do not put the factor $n!$ in (3.46) while others attach this factor to $B_n(x)$.

Exercise 3.6.2. Show that

$$B_n(x) = \frac{1}{n!} \sum_{k=0}^{n} \binom{n}{k} B_k x^{n-k}. \tag{3.47}$$

Use Exercise 3.6.1 to show that $B_0 = 1$, and for $n \geq 1$,

$$B_n = -\frac{1}{n+1} \sum_{k=0}^{n-1} \binom{n+1}{k} B_k. \tag{3.48}$$

The last relation can be written out as a sequence of linear equations:

$$1 + 2B_1 = 0,$$
$$1 + 3B_1 + 3B_2 = 0,$$
$$1 + 4B_1 + 6B_2 + 4B_3 = 0,$$
$$1 + 5B_1 + 10B_2 + 10B_3 + 5B_4 = 0.$$

From here we see that

$$B_1 = -1/2, \quad B_2 = 1/6, \quad B_3 = 0, \quad B_4 = -1/30, \quad B_5 = 0,$$
$$B_6 = 1/42, \quad B_7 = 0, \quad B_8 = -1/30, \ldots.$$

Exercise 3.6.3. Show that

$$\frac{t}{e^t - 1} = \sum_{n=0}^{\infty} \frac{B_n}{n!} t^n. \tag{3.49}$$

(Hint: Write down the power series for $e^t - 1$, multiply the two series, and then equate coefficients.) From this we see that

$$\frac{t}{e^t - 1} + \frac{t}{2} = 1 + \sum_{n=2}^{\infty} \frac{B_n}{n!} t^n. \tag{3.50}$$

The left-hand side can be rewritten as $t(e^t + 1)/2(e^t - 1)$. This is an even function of t. Hence we must have

$$B_{2n+1} = 0 \quad \text{for } n \geq 1. \tag{3.51}$$

What has all this to do with Fourier series? We have seen that the functions $B_n(x), n \geq 2$ are periodic on $\mathbb{R}$ with period 1. The function $B_1(x)$ has the Fourier series

$$B_1(x) = -\frac{1}{\pi} \sum_{n=1}^{\infty} \frac{\sin 2\pi n x}{n}, \quad 0 < x < 1. \tag{3.52}$$

(See (3.20)). Repeated integration shows that for $m \geq 1$,

$$B_{2m}(x) = (-1)^{m-1} \frac{2}{(2\pi)^{2m}} \sum_{n=1}^{\infty} \frac{\cos 2\pi nx}{n^{2m}}, \tag{3.53}$$

$$B_{2m+1}(x) = (-1)^{m-1} \frac{2}{(2\pi)^{2m+1}} \sum_{n=1}^{\infty} \frac{\sin 2\pi nx}{n^{2m+1}} \tag{3.54}$$

for $0 \leq x \leq 1$. (This shows again that $B_{2m+1} = 0$ for all $m \geq 1$.)

The *Riemann zeta function* is the function

$$\zeta(s) = \sum_{n=1}^{\infty} \frac{1}{n^s}, \tag{3.55}$$

a meromorphic function defined on the complex plane. We will be concerned here only with its values at positive integers. We have seen formulas for $\zeta(2)$ and $\zeta(4)$ in (3.39) and (3.44). More generally we have the following.

Theorem 3.6.4. *For every positive integer m the value of $\zeta(2m)$ is a rational multiple of π^{2m} given by the formula*

$$\zeta(2m) = \frac{(-1)^{m-1}(2\pi)^{2m} B_{2m}}{2(2m)!}. \tag{3.56}$$

Exercise 3.6.5. Show that

$$\sum_{n=1}^{\infty} \frac{1}{n^6} = \frac{\pi^6}{945}. \tag{3.57}$$

The result of Theorem 3.6.4 was proved by Euler. The corresponding question about $\zeta(2m+1)$ was left open and has turned out to be very difficult. Not much is known about the values of the zeta function at odd integers. In 1978 R. Apéry showed that $\zeta(3)$ is irrational. Very recently (2000) T. Rivoal has shown that among the values $\zeta(3), \zeta(5), \ldots, \zeta(2n+1)$ at least $\log(n)/3$ must be irrational, and that among the nine values $\zeta(5), \zeta(7), \ldots, \zeta(21)$ at least one is irrational. W. Zudilin has improved this to show that at least one among the four numbers $\zeta(5), \zeta(7), \zeta(9)$ and $\zeta(11)$ is irrational. These methods do not show which of them is irrational.

3.7 $\sin x / x$

The function $\sin x / x$ and the related integral $\text{Si}(x)$ defined as

$$\text{Si}(x) = \int_0^x \frac{\sin t}{t} \, dt$$

appear in several problems. The function $\sin x / x$ is even while $\text{Si}(x)$, called the sine integral, is odd. Their graphs are shown in Figure 6.

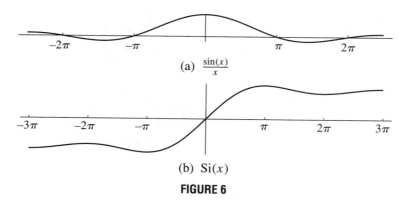

(a) $\dfrac{\sin(x)}{x}$

(b) Si(x)

FIGURE 6

Exercise 3.7.1. Show that the integral

$$\int_0^\infty \left|\frac{\sin t}{t}\right| dt$$

is divergent. (See the discussion in Exercise 2.2.8.)

Exercise 3.7.2. Show that the integral

$$\int_0^\infty \frac{\sin t}{t}\, dt = \lim_{A\to\infty} \int_0^A \frac{\sin t}{t}\, dt \qquad (3.58)$$

is convergent. (Split the integral into two parts

$$\int_0^1 + \int_1^\infty .$$

The first one is clearly finite. Integrate the second by parts. Alternately, represent the integral as an infinite sum with terms

$$\int_{n\pi}^{(n+1)\pi} \sin t/t \, dt.$$

These terms alternate in sign and decrease in absolute value.) Thus we have an example of a function for which the Riemann integral (in the "improper" sense (3.58)) is finite but which is not Lebesgue integrable over $\mathbb{R}$.

The main result of this section is the formula

$$\int_0^\infty \frac{\sin t}{t}\, dt = \frac{\pi}{2}. \qquad (3.59)$$

There are several ways to prove this. We give two proofs related to our discussion of Fourier series. In our discussion of the Dirichlet kernel (Proposition 2.2.2) we have seen that

$$\int_0^\pi \frac{\sin\left((n+1/2)t\right)}{\sin\left(t/2\right)}\, dt = \pi \qquad (3.60)$$

for every positive integer n. Let

$$g(t) = \begin{cases} \dfrac{1}{t/2} - \dfrac{1}{\sin (t/2)}, & 0 < t \le \pi, \\ 0 & t = 0. \end{cases}$$

This is a continuous function (use the L'Hôpital rule). Hence by the Riemann–Lebesgue Lemma (see Exercises 2.3.4 and 2.3.5)

$$\lim_{n\to\infty} \int_0^\pi g(t) \sin \left((n + 1/2)t\right) dt = 0. \tag{3.61}$$

From (3.60) and (3.61) it follows that

$$\lim_{n\to\infty} \int_0^\pi \frac{\sin \left((n + 1/2)t\right)}{t} dt = \frac{\pi}{2}. \tag{3.62}$$

Changing variables by putting $x = (n + 1/2)t$, we get

$$\lim_{n\to\infty} \int_0^{(n+1/2)\pi} \frac{\sin x}{x} dx = \frac{\pi}{2}. \tag{3.63}$$

This is the formula (3.59).

Exercise 3.7.3. The integral I in (3.59) can be written as

$$I = \sum_{n=0}^\infty \int_{n\pi/2}^{(n+1)\pi/2} \frac{\sin t}{t} dt.$$

When $n = 2m$, put $t = m\pi + x$, to get

$$\int_{n\pi/2}^{(n+1)\pi/2} \frac{\sin t}{t} dt = (-1)^m \int_0^{\pi/2} \frac{\sin x}{m\pi + x} dx.$$

When $n = 2m - 1$, put $t = m\pi - x$ to get

$$\int_{n\pi/2}^{(n+1)\pi/2} \frac{\sin t}{t} dt = (-1)^{m-1} \int_0^{\pi/2} \frac{\sin x}{m\pi - x} dx.$$

Hence

$$I = \int_0^{\pi/2} \frac{\sin x}{x} dx + \sum_{m=1}^\infty \int_0^{\pi/2} (-1)^m \left[\frac{1}{x - m\pi} + \frac{1}{x + m\pi}\right] \sin x \, dx.$$

The integrands in the last series are bounded by the terms of a convergent series. Hence

$$I = \int_0^{\pi/2} \sin x \left\{\frac{1}{x} + \sum_{m=1}^\infty (-1)^m \left[\frac{1}{x - m\pi} + \frac{1}{x + m\pi}\right]\right\} dx.$$

The sum in the braces is $1/\sin x$ by Exercise 3.4.7. This gives another proof of (3.59).

Exercise 3.7.4. Use the Taylor expansion

$$\text{Si}(x) = x - \frac{x^3}{3 \cdot 3!} + \frac{x^5}{5 \cdot 5!} \cdots$$

to calculate $\text{Si}(\pi)$ using a pocket calculator. You will see that $\text{Si}(\pi) \simeq 1.852$.

Exercise 3.7.5. Show that for every real number λ,

$$\text{sgn}\,\lambda = \frac{2}{\pi} \int_0^\infty \frac{\sin \lambda t}{t}\,dt.$$

This is an example of an *integral representation* of a function—the function $\text{sgn}\,\lambda$ is expressed as an integral.

Exercise 3.7.6. Show that

$$\int_0^\infty \frac{\sin t \cos t}{t}\,dt = \frac{\pi}{4}.$$

Use this to show that

$$\int_0^\infty \frac{\sin^2 t}{t^2}\,dt = \frac{\pi}{2}.$$

Exercise 3.7.7. Use the formula $\sin x = 2 \sin \frac{x}{2} \cos \frac{x}{2}$ repeatedly and show

$$\frac{\sin x}{x} = \cos \frac{x}{2} \cdot \cos \frac{x}{4} \cdot \cos \frac{x}{8} \cdots.$$

This infinite product formula was discovered by Euler. Using the relation $\cos \frac{x}{2} = \sqrt{(1 + \cos x)/2}$ and choosing $x = \pi/2$, obtain from this the formula

$$\frac{2}{\pi} = \frac{\sqrt{2}}{2} \cdot \frac{\sqrt{2 + \sqrt{2}}}{2} \cdot \frac{\sqrt{2 + \sqrt{2 + \sqrt{2}}}}{2} \cdots.$$

This is called *Viète's product formula*. Note that this formula allows us to obtain an approximate value of π by repeatedly using four basic operations of arithmetic (addition, multiplication, division, and square root extraction) all applied to a single number 2. Use a pocket calculator to see what approximate value of π is obtained by taking the first ten terms of this product.

3.8 The Gibbs phenomenon

We have seen that if f is piecewise C^1, then its Fourier series converges at each point where f is continuous, and at a point of discontinuity the series converges to the average value of f at that point. However, in a neighbourhood of a discontinuity the convergence of the series is not uniform and the partial sums of the series always overestimate the function by about 18%. This is called the *Gibbs phenomenon*.

Consider the function of the Example 3.3.5 and its odd extension having the Fourier sine series

$$\frac{2}{\pi} \sum_{n=1}^{\infty} (-1)^{n+1} \frac{1}{n} \sin n\pi x.$$

For each N consider the partial sums $S_N(f; x)$ at the points $x = 1 - 1/N$ and $x = -1 + 1/N$. We have

$$S_N\left(f; 1 - \frac{1}{N}\right) = \frac{2}{\pi} \sum_{n=1}^{N} \frac{(-1)^{n+1}}{n} \sin\left(n\pi - \frac{n\pi}{N}\right)$$

$$= \frac{2}{\pi} \sum_{n=1}^{N} \frac{1}{n} \sin \frac{n\pi}{N}.$$

$$= \frac{2}{\pi} \sum_{n=1}^{N} \frac{\sin(n\pi/N)}{n\pi/N} \cdot \frac{\pi}{N}.$$

This last sum is a Riemann sum for the integral

$$\int_0^{\pi} \frac{\sin t}{t} \, dt.$$

So we have, using Exercise 3.7.4,

$$\lim_{N \to \infty} S_N\left(f; 1 - \frac{1}{N}\right) = \frac{2}{\pi} \int_0^{\pi} \frac{\sin t}{t} \, dt > 1.17.$$

By the same argument one can see that

$$\lim_{N \to -\infty} S_N\left(f; -1 + \frac{1}{N}\right) < -1.17.$$

The values of f at the points $x = 1^-$ and $x = -1^+$ are 1 and -1 respectively. However the sums of the Fourier series *overshoot* these limits.

If such calculations trouble you, you might find inspiration from the history of the discovery of the Gibbs phenomenon.

You will have heard of A. Michelson, one of the greatest experimental physicists and an extraordinary designer and builder of equipment to do the experiments. He is known for his accurate measurement of the speed of light and for the Michelson–Morley experiments to detect ether. He also designed and built a machine to calculate Fourier series. To test his machine he fed into it the first 80 Fourier coefficients of the function f we have used above. He was surprised that he did not get back the original function but the machine seemed to add two little peaks near the discontinuities roughly as shown below.

The effect of increasing n was to move these peaks closer and closer to ± 1 but not to diminish their size as Michelson expected. After verifying his observation by hand calculations he wrote a letter to the magazine *Nature* in 1898 expressing his doubts that "a real discontinuity [in f] can replace a sum of continuous curves $[S_n(f; \theta)]$." Gibbs—another great physicist who was one of the founders of modern thermodynamics—replied to the letter and clarified the matter.

Exercise 3.8.1. You can clarify your understanding of uniform convergence by reconciling the Gibbs phenomenon with Theorem 2.3.19 on convergence of Fourier series of piecewise C^1 functions.

Exercise 3.8.2. We have demonstrated the existence of the Gibbs phenomenon for a particular function. Show that this must occur near the jumps of any piecewise C^1 function.

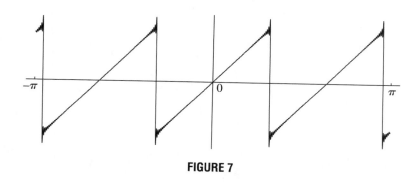

FIGURE 7

3.9 Exercises

The common theme of the exercises in this section is *positivity*. The numbers a_n, b_n and c_n in Exercises 3.9.1–3.9.4 are the Fourier coefficients associated with f as in (3.5) and (3.6). (The index n runs over all nonnegative integers for a_n, over all positive integers for b_n, and over all integers for c_n.)

Exercise 3.9.1. Let $f(t) \geq 0$. Show that for all n, $|c_n| \leq c_0$, $|a_n| \leq a_0$, and $|b_n| \leq a_0$.

Exercise 3.9.2. Let f be an odd function such that $f(t) \geq 0$ for $0 < t < \pi$. Show that $|b_n| \leq nb_1$ for all n.

Exercise 3.9.3. Let f be a monotonically decreasing function on the interval $(0, \pi)$. Show that $b_n \geq 0$ for all n.

Exercise 3.9.4. Let f be a convex function on $(0, 2\pi)$. Show that $a_n \geq 0$ for $n = 1, 2, \ldots$.

A doubly infinite sequence $\{c_n\}$ of complex numbers is said to be a *positive definite sequence* if

$$\sum_{0 \leq r,s \leq N-1} c_{r-s} z_r \bar{z}_s \geq 0 \tag{3.64}$$

for all positive integers N, and for all complex numbers $z_0, z_1, \ldots, z_{N-1}$. This is equivalent to saying that for every N the $N \times N$ matrix

$$\begin{bmatrix} c_0 & c_{-1} & c_{-2} & \cdots & c_{-(N-1)} \\ c_1 & c_0 & c_{-1} & \cdots & \cdot \\ c_2 & c_1 & c_0 & \cdots & \cdot \\ \cdot & \cdot & \cdot & \cdot & \cdot \\ c_{N-1} & c_{N-2} & \cdot & \cdot & c_0 \end{bmatrix} \tag{3.65}$$

is positive (semi) definite. (This matrix has a special form; each of its diagonals is constant. Such a matrix is called a Toeplitz matrix.) Clearly, if this matrix is positive definite, then

$$c_{-n} = \bar{c}_n, \qquad c_0 \geq 0, \qquad |c_n| \leq c_0.$$

Exercise 3.9.5. Let $f \in C(T)$ and let $\{c_n\}$ be the Fourier coefficients of f.

(i) If $f(t) \geq 0$ show that the sequence $\{c_n\}$ is positive definite.

(ii) Conversely, suppose $\{c_n\}$ is positive definite. Show that for all N

$$\sum_{0 \leq r,s \leq N-1} c_{r-s} e^{i(r-s)t} \geq 0.$$

This implies that the Fejér sums

$$\sigma_N(f; t) = \sum_{n=-(N-1)}^{N-1} \left(1 - \frac{|n|}{N}\right) c_n e^{int} \tag{3.66}$$

are nonnegative. Hence, by Fejér's Theorem (2.2.5), $f(t) \geq 0$.

This statement is a part of a general theorem called *Herglotz's Theorem* which says that every positive definite sequence is the sequence of Fourier coefficients associated with a positive measure μ; i.e.,

$$c_n = \int_{-\pi}^{\pi} e^{-int} d\mu(t). \tag{3.67}$$

(This can be proved along the lines of Exercise 3.9.5. Given c_n, let $f_N(t)$ be the sequence of functions defined by the expression on the right-hand side of (3.66). Then $\{f_N\}$ is a family of nonnegative continuous functions with the property

$$\frac{1}{2\pi} \int_{-\pi}^{\pi} f_N(t) \, dt = c_0.$$

This is a kind of compactness condition from which we can derive (3.67) using standard theorems of Functional Analysis.)

Exercise 3.9.6. A linear operator A on the space $C(T)$ is called *positive* if it maps non-negative functions to nonnegative functions. (The elements of $C(T)$ are complex-valued continuous functions on T). Let g_0, g_1, g_2 be the three functions

$$g_0(t) = 1, \qquad g_1(t) = \sin t, \qquad g_2(t) = \cos t.$$

Korovkin's Theorem is the statement: If $\{A_n\}$ is a sequence of positive linear operators on $C(T)$ such that as $n \to \infty$,

$$A_n g_j \to g_j \quad \text{uniformly, for} \quad j = 0, 1, 2,$$

then

$$A_n f \to f \quad \text{uniformly for all} \quad f \in C(T).$$

A proof of this theorem is sketched below. Fill in the details.

(i) It is enough to prove the assertion for all real-valued functions f.
(ii) Fix $s \in T$ and let $g_s(t) = \sin^2[(t - s)/2]$. Note that $g_s(t) = g_0(t) - g_1(s)g_1(t) - g_2(s)g_2(t)$.
(iii) Let $M = \sup |f|$. Let $\varepsilon > 0$, and let δ be such that $|f(t) - f(s)| < \varepsilon$ whenever $|t - s| < \delta$. Note that $g_s(t) \geq 4\delta^2/\pi^2$, whenever $|t - s| \geq \delta$. Thus

$$|f(t) - f(s)| \leq \varepsilon + \frac{M\pi^2}{2\delta^2} g_s(t).$$

(iv) Put $K = M\pi^2/2\delta^2$, a constant that depends on f and ε. The last inequality can be written as

$$-\varepsilon g_0 - K g_s \leq f - f(s)g_0 \leq \varepsilon g_0 + K g_s,$$

an inequality between functions (in the pointwise sense). Since the operators A_n are linear and order preserving, this gives

$$|(A_n f)(t) - f(s)(A_n g_0)(t)| \leq \varepsilon (A_n g_0)(t) + K(A_n g_s)(t) \tag{3.68}$$

for all t.
(v) Choose $t = s$ in (3.68) and let $n \to \infty$. Note $(A_n g_0)(s) \to 1$, and by the observation in (ii), $(A_n g_s)(s) \to 0$. The convergence is uniform over s. So from (3.68) we see that $(A_n f)(s) \to f(s)$ uniformly.

Exercise 3.9.7. Use Korovkin's Theorem to give another proof of Fejér's Theorem. (Hint: The Fejér sums $A_n f = \sigma_n(f)$ define positive operators on $C(T)$.)

Exercise 3.9.8. There is a Korovkin Theorem for the space $C[a, b]$ where $[a, b]$ is any interval. The three functions g_0, g_1, g_2 now are $g_o(t) = 1, g_1(t) = t, g_2(t) = t^2$. Formulate and prove the assertion in this case.

3.10 A historical digression

The series (3.39) is a famous one. The mathematician brothers Jakob and Johann Bernoulli had discovered in 1689 a simple argument to show that the harmonic series $\sum 1/n$ is divergent. This was contrary to the belief, held earlier, that a series $\sum x_n$ with positive terms decreasing to zero converges. Among other series they tried to sum was the series $\sum 1/n^2$. They could see that this series is convergent to a sum less than 2 but did not succeed in finding the sum. This problem became famous as the "Basel Problem" after the Swiss town Basel where the Bernoullis lived.

Solving the Basel problem was one of the early triumphs of L. Euler (1707–1783) who went on to become one of the most prolific and versatile mathematicians of all time. Euler was a student of Jakob and Johann, and a friend of Johann's sons Daniel and Nicolaus (all mathematicians—there were more in the family). It is worthwhile to recall, very briefly, some of the stages in Euler's attack on this problem as they are illustrative of a great scientist's working.

In 1731 Euler proved the formula

$$\zeta(2) = (\log 2)^2 + 2 \sum_{n=1}^{\infty} \frac{1}{2^n n^2},$$

using calculus and infinite series expansions of $\log(1 - x)$. In this one can substitute

$$\log 2 = \sum_{n=1}^{\infty} \frac{1}{n 2^n}.$$

The factors 2^n occurring here make the series converge very rapidly. Using this he found that

$$\zeta(2) = 1.644934\ldots.$$

This number did not look familiar. More calculations of this kind followed. The proof of (3.39), however, came through another route: Newton's theorem on roots of polynomials generalized to "infinite degree polynomials."

Let $p(x)$ be a polynomial of degree n and suppose $p(0) = 1$. Suppose we know the n numbers $\alpha_1, \alpha_2, \ldots, \alpha_n$ are roots of p. Then we can write

$$p(x) = \left(1 - \frac{x}{\alpha_1}\right)\left(1 - \frac{x}{\alpha_2}\right) \cdots \left(1 - \frac{x}{\alpha_n}\right). \tag{3.69}$$

Euler considered the real function

$$f(x) = \frac{\sin x}{x} = 1 - \frac{x^2}{3!} + \frac{x^5}{5!} \cdots, \tag{3.70}$$

noticed $f(0) = 1$, and $f(x) = 0$ precisely when $x = n\pi, n \in Z$. In analogy with (3.69) he factored (3.70) as

$$f(x) = \left[\left(1 - \frac{x}{\pi}\right)\left(1 + \frac{x}{\pi}\right)\right]\left[\left(1 - \frac{x}{2\pi}\right)\left(1 + \frac{x}{2\pi}\right)\right]\cdots$$

$$= \left(1 - \frac{x^2}{\pi^2}\right)\left(1 - \frac{x^2}{4\pi^2}\right)\cdots.$$

He then multiplied out this infinite product to write

$$f(x) = 1 - \left(\frac{1}{\pi^2} + \frac{1}{4\pi^2} + \frac{1}{9\pi^2} + \cdots\right)x^2 + (\cdots)x^4 + \cdots. \qquad (3.71)$$

Now comparing coefficients of x^2 in (3.70) and (3.71) he saw that

$$\frac{1}{3!} = \frac{1}{\pi^2}\sum\frac{1}{n^2}.$$

This is the remarkable formula (3.39). The answer confirmed the numerical calculations made earlier by Euler and others.

Several questions arise when one sees such a calculation. Can one do to an infinite series whatever one does to a polynomial? One knows all the *real* roots of $f(x)$, but are there other roots? What is the meaning of an infinite product? The function $g(x) = e^x f(x)$ has the same roots as $f(x)$. Can both be represented by the same formula?

Euler certainly was aware of all these problems and spent several years resolving them. This led to a more rigorous proof of the product expansion (3.33). These days infinite product expansions for entire functions (and meromorphic functions) are routinely used in analysis. (See the *Weierstrass Factorisation Theorem* and the *Mittag-Leffler Theorem* in L. V. Ahlfors, *Complex Analysis*.)

Euler found great delight in coaxing out of equation (3.71) more and more complicated sums. The connection with Bernoulli numbers (introduced by Jakob Bernoulli in his famous work *Ars Conjectandi*) became apparent and he proved a general formula for $\zeta(2k)$ in terms of these numbers.

You may enjoy reading two books by W. Dunham. Chapters 8 and 9 of his book *Journey Through Genius* are titled *The Bernoullis and the Harmonic Series*, and *The Extraordinary Sums of Leonhard Euler*. The other book *Euler: The Master of Us All* is a short elementary account of some of Euler's work. Chapter 3 of A. Weil, *Number Theory: An Approach Through History from Hammurapi to Legendre* is devoted to Euler. This book by a great twentieth century mathematician is an excellent historical introduction to a substantial part of mathematics.

The proof of (3.39) that we have given is one of several known now. Like Euler's proof this uses a general method that leads to several other formulas. A more special proof given by Euler is outlined in the following exercise:

Exercise 3.10.1.

(i) Show that

$$\frac{1}{2}(\sin^{-1} x)^2 = \int_0^x \frac{\sin^{-1} t}{\sqrt{1 - t^2}} \, dt.$$

(ii) Use the integral

$$\sin^{-1} x = \int_0^x (1 - t^2)^{-1/2} \, dt$$

to expand $\sin^{-1} x$ in a power series. Substitute this in the integral in (i) and integrate term by term. Use the recurrence relation

$$\int_0^1 \frac{t^{n+2}}{\sqrt{1 - t^2}} \, dt = \frac{n + 1}{n + 2} \int_0^1 \frac{t^n}{\sqrt{1 - t^2}} \, dt$$

starting with

$$\int_0^1 \frac{t}{\sqrt{1 - t^2}} \, dt = 1.$$

(iii) The result of this calculation is a power series expansion for the function $\frac{1}{2}(\sin^{-1} x)^2$. Put $x = 1$ in this formula to obtain the sum (3.38).

Exercise 3.10.2. Show that

$$\sum_{n=1}^{\infty} \frac{1}{n^2} = 3 \sum_{n=1}^{\infty} \frac{(n - 1)^2}{(2n)!}.$$

[Hint: Use the power series for $(\sin^{-1} x)^2$.]

One of the reasons for interest in the zeta function is the intimate connection it has with prime numbers.

Exercise 3.10.3. For $s > 1$, let

$$\zeta(s) = \sum_{n=1}^{\infty} \frac{1}{n^s}.$$

Show that

$$\zeta(s) = \prod_p \frac{1}{1 - p^{-s}},$$

where, in the product on the right, p varies over all prime numbers. Among other things this shows that there are infinitely many prime numbers. This formula was discovered by Euler.

Let us demonstrate an interesting application of it that links prime factors, probability, and the number π. A naive and intuitive idea of probability is adequate for our discussion.

A natural number picked "at random" is as likely to be even as odd. We express this by saying that the probability of a natural number being even (or odd) is $1/2$. In the same way, the probability of a natural number picked at random being a multiple of k is $1/k$ (as such multiples occur at jumps of length k in the sequence of natural numbers).

Pick up a natural number n at random and factor it into primes. What is the probability that no prime is repeated in this factoring? This happens if n is not a multiple of p^2 for any prime p. As we have seen, the probability of this for any given prime p is $1 - 1/p^2$. The probability of all these "events" happening simultaneously is the product of these individual probabilities, i.e., $\prod_p (1 - 1/p^2)$. We have seen that this product is equal to

$$\frac{1}{\zeta(2)} = \frac{6}{\pi^2}.$$

Exercise 3.10.4. Think of a plausible argument that shows that the probability that two natural numbers picked at random are coprime is $6/\pi^2$. Generalise this statement to k numbers.

The formulas of Viète and Wallis (Exercises 3.7.7 and 3.5.3) were discovered in 1593 and 1655, respectively. Thus they preceded the work of Euler (1707–1783). Of course, the arguments of Viète and Wallis were different from the ones given here. Since both formulas give expressions for $2/\pi$, one may wonder whether there is a single expression that unites them. There is one, and it was found quite recently. The curious reader should see the paper T. Osler, The union of Vièta's and Wallis's product for π, *American Mathematical Monthly*, Volume 106, October, 1999, pp. 774–776.

A final tidbit. The symbol π seems to have been first used by William Jones in 1706. For thirty years it was not used again till Euler used it in a treatise in 1736. It won general acceptance when Euler used it again in his famous book *Introduction to the Analysis of the Infinite* published in 1748. There are a few books devoted exclusively to the number π. A recent one is J. Arndt and C. Haenel, π *Unleashed*, Springer, 2001.

4

Convergence in L_2 and L_1

In Chapter 2 we saw that the Fourier series of a continuous function on T may not converge at every point. But under weaker notions of convergence (like Cesàro convergence) the series does converge. Another notion of convergence is that of convergence in the mean square or L_2 convergence. However, this notion is in some sense even more natural than pointwise convergence for Fourier series. We will now study this and the related L_1 convergence. We assume that the reader is familiar with basic properties of Hilbert space and the spaces L_1 and L_2. Some of them are quickly recalled below in the form of exercises.

4.1 L_2 convergence of Fourier series

Let $\mathcal{H}$ be a Hilbert space with inner product $\langle ., . \rangle$ and norm $\| \cdot \|$. Special examples of interest to us are:

1. The space l_2 of all sequences of complex numbers $x = (x_1, x_2, \ldots)$ such that

$$\sum |x_n|^2 < \infty.$$

The inner product in this space is defined as

$$\langle x, y \rangle = \sum_{n=1}^{\infty} x_n \bar{y}_n$$

and the associated norm is

$$\|x\|_2 = \left(\sum_{n=1}^{\infty} |x_n|^2 \right)^{1/2}.$$

We need also the space of *doubly infinite* sequences $\{x_n\}_{n=-\infty}^{\infty}$ satisfying

$$\sum_{n=-\infty}^{\infty} |x_n|^2 < \infty.$$

We denote this space by $l_2(Z)$. It is easy to see that l_2 and $l_2(Z)$ are isomorphic Hilbert spaces.

2. For any bounded interval I on the real line, the space $L_2(I)$ consists of all Lebesgue measurable functions f on I such that $\int_I |f|^2 < \infty$. The inner product on this space is defined as

$$\langle f, g \rangle = \frac{1}{|I|} \int_I f(x)\, \overline{g(x)}\, dx,$$

where $|I|$ denotes the length of I. The associated norm is

$$\|f\|_2 = \left(\frac{1}{|I|} \int_I |f(x)|^2\, dx \right)^{1/2}.$$

(The factor $1/|I|$ is inserted to make some calculations simpler.)

3. The space $L_2(\mathbb{R})$ consists of all Lebesgue measurable functions on the real line $\mathbb{R}$ such that $\int_{\mathbb{R}} |f|^2 < \infty$. Here the inner product and the norm are given as

$$\langle f, g \rangle = \int_{\mathbb{R}} f(x)\overline{g(x)}\, dx,$$

$$\|f\|_2 = \left(\int_{\mathbb{R}} |f(x)|^2\, dx \right)^{1/2}.$$

Of course, when we talk of measurable functions, we identify functions that are equal almost everywhere.

The space $L_2[-\pi, \pi]$ can be identified with $L_2(T)$.

We are interested also in the space $L_1(I)$ which consists of all integrable functions on I, i.e.,

$$L_1(I) = \left\{ f : \int_I |f(x)|\, dx < \infty \right\}.$$

When I is a bounded interval, we define, for $f \in L_1(I)$,

$$\|f\|_1 = \frac{1}{|I|} \int_I |f(x)|\, dx.$$

For f in the space $L_1(\mathbb{R})$ we define

$$\|f\|_1 = \int_{\mathbb{R}} |f(x)|\, dx.$$

The spaces $L_1(I)$ and $L_1(\mathbb{R})$ are Banach spaces with this norm. Neither of them is a Hilbert space. (You can prove this by showing that the norm does not satisfy the "parallelogram law" which a Hilbert space norm must satisfy).

Exercise 4.1.1. Show that the norm and the inner product in any Hilbert space satisfy the *Cauchy–Schwarz inequality:*

$$|\langle x, y \rangle| \leq \|x\| \|y\|.$$

Exercise 4.1.2. Use this inequality to show that for every bounded interval I,

$$L_2(I) \subset L_1(I).$$

Show that this inclusion is proper.

Exercise 4.1.3. Show that neither of $L_2(\mathbb{R})$ and $L_1(\mathbb{R})$ is contained in the other.

Exercise 4.1.4. Show that continuous functions are a dense subset of L_2 and of L_1. Functions of class C^∞ also are dense in L_2 and in L_1.

Recall that a sequence $\{e_n\}$ in a Hilbert space $\mathcal{H}$ is called *orthogonal* if $\langle e_n, e_m \rangle = 0$ for $n \neq m$, and *orthonormal* if in addition $\|e_n\| = 1$. Given any vector x in $\mathcal{H}$ let $\langle x, e_n \rangle = a_n$. The numbers a_n are called the *Fourier coefficients of x with respect to the orthonormal system* $\{e_n\}$.

Exercise 4.1.5. Show that in the Hilbert space $L_2(T)$ the sequence e_n, defined as

$$e_n(x) = e^{inx}, \qquad n \in Z,$$

is an orthonormal sequence. Note that here the Fourier coefficients are

$$\langle f, e_n \rangle = \frac{1}{2\pi} \int_{-\pi}^{\pi} f(x) e^{-inx} \, dx = \hat{f}(n).$$

So the general results on orthonormal sets in Hilbert spaces are applicable here.

Exercise 4.1.6. (Bessel's inequality) Prove that if $\{e_n\}$ is any orthonormal system in a Hilbert space $\mathcal{H}$ and a_n the Fourier coefficients of a vector x with respect to this system, then

$$\sum_n |a_n|^2 \leq \|x\|^2.$$

Therefore, the sequence $\{a_n\}$ belongs to the space l_2. For $f \in L_2(T)$ this gives

$$\sum_n |\hat{f}(n)|^2 \leq \|f\|^2.$$

Exercise 4.1.7. Let a_n be the Fourier coefficients of a vector x with respect to an orthonormal system $\{e_n\}$. Let b_n be any other sequence of numbers. Show that

$$\left\| x - \sum_{n=1}^{N} a_n e_n \right\| \leq \left\| x - \sum_{n=1}^{N} b_n e_n \right\|$$

for all integers N. Further, this inequality is an equality if and only if $a_n = b_n$ for all $n = 1, 2, \ldots, N$. For the space $L_2(T)$ this means that among all trigonometric polynomials of degree N, the partial sum $S_N(f)$ is the closest approximation to a given function f in the norm of $L_2(T)$.

An orthonormal system $\{e_n\}$ in a Hilbert space $\mathcal{H}$ is called *complete* if finite linear combinations of $\{e_n\}$ are dense in $\mathcal{H}$, i.e., given $x \in \mathcal{H}$ and $\varepsilon > 0$ there exists an integer N and numbers $a'_1 \ldots, a'_N$ such that

$$\left\| x - \sum_{n=1}^{N} a'_n e_n \right\| < \varepsilon.$$

Note that, by Exercise 4.1.7, the best choice for such numbers, if they exist, is the Fourier coefficients a_n. A complete orthonormal system is called a *basis* for $\mathcal{H}$. The space $\mathcal{H}$ is *separable* if and only if it has a countable basis.

Exercise 4.1.8. (Plancherel's Theorem) Show that an orthonormal system $\{e_n\}$ is a basis in $\mathcal{H}$ if and only if the Fourier coefficients a_n of every vector x with respect to e_n satisfy the equality

$$\sum_n |a_n|^2 = \|x\|^2,$$

i.e., Bessel's inequality becomes an equality.

Theorem 4.1.9. *The system $e_n(x) = e^{inx}$ is a basis for $L_2(T)$.*

Proof Let $f \in L_2(T)$. By Exercise 4.1.4, for each $\varepsilon > 0$, there exists $g \in C^1(T)$ such that $\|f - g\|_2 \le \frac{\varepsilon}{2}$. By Dirichlet's Theorem there exists an N such that

$$\sup |g(x) - S_N(g; x)| < \frac{\varepsilon}{2}.$$

Denote $S_N(g; x)$ by S_N. Then

$$\|g - S_N\|_2 = \left[\frac{1}{2\pi} \int_{-\pi}^{\pi} |g(x) - S_N(g; x)|^2 \, dx \right]^{1/2} \le \frac{\varepsilon}{2}.$$

Hence

$$\|f - S_N\|_2 \le \varepsilon.$$

But S_N is a linear combination of the functions e_n. ∎

Exercise 4.1.10. Show that for $f \in L_2(T)$,

$$\lim_{N \to \infty} \left\| f - \sum_{n=-N}^{N} \hat{f}(n) e_n \right\|_2 = 0,$$

$$\|f\|_2^2 = \sum_{n=-\infty}^{\infty} |\hat{f}(n)|^2.$$

Thus the Fourier series $\sum \hat{f}(n)e^{inx}$ converges to f in the sense of convergence with respect to the norm of $L_2(T)$. In particular, this is true for all continuous functions on T. This explains our remark at the beginning of the chapter.

Exercise 4.1.11. Let $\{a_n\}$ be any sequence in l_2. Let $\mathcal{H}$ be a Hilbert space with an orthonormal basis $\{e_n\}$. Show that there exists a vector x in $\mathcal{H}$ whose Fourier coefficients with respect to e_n are a_n. Hint: Show that the series $\sum a_n e_n$ converges in $\mathcal{H}$.

In particular, given any sequence $\{a_n\}_{n=-\infty}^{\infty}$ in $l_2(Z)$ we can find a function f in $L_2(T)$ such that

$$\hat{f}(n) = a_n.$$

Exercise 4.1.12. (Riesz–Fischer Theorem) Show that the spaces $L_2(T)$ and l_2 are isomorphic: the map that sends an element f of $L_2(T)$ to the sequence consisting of its Fourier coefficients is an isomorphism. (This is a very important fact useful in several contexts. For example, in Quantum Mechanics the equivalence between the "Wave Mechanics" and the "Matrix Mechanics" approaches is based on this fact. See the classic book, J. von Neumann, *Mathematical Foundations of Quantum Mechanics*.)

Exercise 4.1.13. (Parseval's Relations) Show that if $f, g \in L_2(T)$, then

$$\frac{1}{2\pi} \int_{-\pi}^{\pi} f(x)\overline{g}(x)\,dx = \sum_{n=-\infty}^{\infty} \hat{f}(n)\overline{\hat{g}(n)}.$$

Prove a general version of this for any Hilbert space.

Exercise 4.1.14. Use the Weierstrass approximation theorem (Exercise 3.1.4) and Exercise 4.1.4 to show that the polynomial functions are dense in the Hilbert space $L_2[-1, 1]$. Note that a polynomial is a finite linear combination of the functions $x^n, n \geq 0$. Obtain an orthonormal basis of $L_2[-1, 1]$ by applying the familiar Gram–Schmidt process to the functions x^n. The functions you get as a result are, up to constant multiples,

$$P_0(x) = 1,$$

$$P_1(x) = x,$$

$$P_2(x) = \frac{3}{2}x^2 - \frac{1}{2},$$

$$P_3(x) = \frac{5}{2}x^3 - \frac{3}{2}x,$$

$$\vdots$$

$$P_n(x) = \frac{1}{2^n n!}\frac{d^n}{dx^n}(x^2 - 1)^n.$$

These are called the *Legendre Polynomials*. Every function in $L_2[-1, 1]$ can be expanded with respect to the basis formed by these functions. This is called the *Fourier–Legendre series.* (There are several other orthonormal systems for the space L_2 that are useful in the study of differential equations of physics).

Exercise 4.1.15. Let $\{e_n\}$ be an orthonormal system in a Hilbert space $\mathcal{H}$. Show that the following conditions are equivalent:

  (i) $\{e_n\}$ is a basis,
 (ii) if $x \in \mathcal{H}$ is such that $\langle x, e_n \rangle = 0$ for all n, then $x = 0$,
(iii) $\|x\|^2 = \sum_n |\langle x, e_n \rangle|^2$ for all $x \in \mathcal{H}$.

These results are very useful in different contexts. Here are some examples.

Exercise 4.1.16. Let $f \in C^1(T)$. Recall that then $in \hat{f}(n) = \widehat{f'}(n)$. (See Exercise 2.3.11). Use this and the Cauchy–Schwarz and the Bessel inequalities to show that, for $0 < N < N' < \infty$,

$$|S_N(f; \theta) - S_{N'}(f; \theta)| \leq \sum_{|n|>N} |\hat{f}(n)|$$

$$= \sum_{|n|>N} \frac{1}{|n|} |\widehat{f'}(n)|$$

$$\leq \frac{C}{N^{1/2}} \|f'\|_2.$$

This shows that $S_N(f; \theta)$ converges to $f(\theta)$ uniformly at the rate $1/N^{1/2}$ as $N \to \infty$.

By the same argument, show that if $f \in C^k(T)$, then $S_N(f; \theta)$ converges to $f(\theta)$ uniformly at the rate $1/N^{k-1/2}$.

Exercise 4.1.17. In Exercise 2.3.12 we saw that if $f \in C^k(T)$, then $\hat{f}(n) = O(1/n^k)$. Show that if $f \in L_1(T)$ and $\hat{f}(n) = O(1/n^k)$ for some integer $k > 2$, then f is $k - 2$ times continuously differentiable. Show that no more can be concluded.
Hint: $\sum \hat{f}(n) in e^{inx}$ converges uniformly.

Exercise 4.1.18. A sequence $\{x_n\}_{n=-\infty}^{\infty}$ is called *rapidly decreasing* if $|n|^k x_n$ converges to zero as $|n| \to \infty$, for all $k \geq 0$. Construct an example of such a sequence. Show that $f \in C^\infty(T)$ if and only if the sequence $\hat{f}(n)$ is rapidly decreasing.

Exercise 4.1.19. Let f have the Fourier series

$$\frac{a_0}{2} + \sum_{n=1}^{\infty} (a_n \cos n\theta + b_n \sin n\theta).$$

(See (3.5) in Chapter 3). Show that

$$\frac{1}{\pi} \int_{-\pi}^{\pi} |f|^2 = \frac{a_0^2}{2} + \sum_{n=1}^{\infty} (a_n^2 + b_n^2).$$

Derive a corresponding result when f is a function with an arbitrary period.

Exercise 4.1.20. Use this version of Plancherel's Theorem to prove

$$\sum_{n=1}^{\infty} \frac{1}{n^2} = \frac{\pi^2}{6}.$$

Hint: Consider the Fourier series of the function in Example 3.3.5.

Exercise 4.1.21. Derive more formulas of this kind by applying Plancherel's Theorem to functions studied in Section 3.3.

Exercise 4.1.22. Let f be a C^1 function on $[0, \pi]$ with $f(0) = f(\pi) = 0$. Show that

$$\int_0^{\pi} |f|^2 \le \int_0^{\pi} |f'|^2.$$

By a change of variables, show that if f is a C^1 function on any interval $[a, b]$ with $f(a) = f(b) = 0$, then

$$\int_a^b |f|^2 \le \frac{(b-a)^2}{\pi^2} \int_a^b |f'|^2.$$

This is called *Wirtinger's inequality*.

Show that these inequalities are sharp. Hint: Let $f(x) = \sin x$.

4.2 Fourier coefficients of L_1 functions

The Fourier coefficients

$$\hat{f}(n) = \frac{1}{2\pi} \int_{-\pi}^{\pi} f(t) e^{-int} \, dt$$

are defined whenever the integral on the right-hand side makes sense. This is so whenever $f \in L_1(T)$. The spaces $C(T)$ and $L_2(T)$ we have considered earlier are subspaces of $L_1(T)$. For $L_2(T)$ we have obtained a complete characterization of the Fourier coefficients: we know that $f \in L_2(T)$ iff the sequence $\hat{f}(n)$ is in l_2, and further the map taking f to the sequence $\{\hat{f}(n)\}$ is an isomorphism of Hilbert spaces. For $L_1(T)$ the situation is far less satisfactory, as we will see in this section. One of the more attractive features of $L_1(T)$ is that it is a "convolution algebra" and there is a very neat theorem relating the Fourier coefficients of $f * g$ with those of f and g. Many of our results for L_1 can be deduced from the ones we have already derived for continuous functions.

For $f \in C(T)$ define

$$\|f\|_{\infty} = \sup_{t \in T} |f(t)|.$$

Then note

$$\|f\|_1 = \frac{1}{2\pi} \int_{-\pi}^{\pi} |f(t)| \, dt \le \|f\|_{\infty}. \tag{4.1}$$

So if $f_n \to f$ in the space $C(T)$, i.e., if the sequence f_n of continuous functions converges uniformly to f, then $f_n \to f$ in $L_1(T)$ also, i.e., $\|f_n - f\|_1 \to 0$.

For $f, g \in L_1(T)$ define, as before,

$$(f * g)(t) = \int_{-\pi}^{\pi} f(t - x)g(x)\,dx.$$

A routine argument using Fubini's Theorem shows that $f * g$ is well defined for almost all t, and is in $L_1(T)$. It is perhaps worthwhile to go through this argument in detail, though it is somewhat dull.

Let I, J be any two (finite or infinite) intervals and let $I \times J$ be their product. The theorems of Fubini and Tonelli relate the double integral of a measurable function $f(x, y)$ with respect to the product measure on $I \times J$ to iterated one-variable integrals.

Fubini's Theorem. *Let $f \in L_1(I \times J)$, i.e., let*

$$\iint_{I \times J} |f(x, y)|\,dx\,dy < \infty.$$

Then for almost all $x \in I$,

$$\int_J |f(x, y)|\,dy < \infty,$$

i.e., the function $f_x(y) = f(x, y)$ is an integrable function of y for almost all x. Further, the function

$$\int_J f(x, y)\,dy$$

is an integrable function of x, and

$$\iint_{I \times J} f(x, y)\,dx\,dy = \int_I \left[\int_J f(x, y)\,dy \right] dx.$$

A corresponding statement with the roles of x and y interchanged is true.

Tonelli's Theorem. *Let f be a nonnegative measurable function on $I \times J$. If one of the integrals*

$$\int_I \left[\int_J f(x, y)\,dy \right] dx,$$

$$\int_J \left[\int_I f(x, y)\,dx \right] dy$$

is finite, then so is the other, and $f \in L_1(I \times J)$.

Hence (by Fubini's Theorem)

$$\iint_{I \times J} f(x, y)\, dx\, dy = \int_I \left[\int_J f(x, y)\, dy \right] dx$$

$$= \int_J \left[\int_I f(x, y)\, dx \right] dy.$$

(Fubini's Theorem says that if $f(x, y)$ is an integrable function on $I \times J$, then its integral can be evaluated in either of the three ways—as an integral with respect to the area measure or as an iterated integral in two different ways. Tonelli's Theorem says that if f is nonnegative, then even the hypothesis of integrability is unnecessary. The three integrals are either all finite or all infinite.)

Let us return to convolutions now.

Theorem 4.2.1. *Let $f, g, \in L_1(T)$. Then the integral defining $(f * g)(t)$ exists absolutely for almost all t, is in $L_1(T)$, and*

$$\| f * g \|_1 \leq 2\pi \| f \|_1 \| g \|_1. \tag{4.2}$$

Proof Since f is a periodic function,

$$\int_{-\pi}^{\pi} |f(t - x)|\, dt = \int_{-\pi}^{\pi} |f(t)|\, dt = 2\pi \| f \|_1$$

for all $x \in T$. Hence

$$\int_{-\pi}^{\pi} |f(t - x)g(x)|\, dt = |g(x)| \int_{-\pi}^{\pi} |f(t - x)|\, dt = |g(x)| 2\pi \| f \|_1,$$

and therefore,

$$\int_{-\pi}^{\pi} \left[\int_{-\pi}^{\pi} |f(t - x)g(x)|\, dt \right] dx = 4\pi^2 \| f \|_1 \, \| g \|_1.$$

So, by Tonelli's theorem $f(t - x)g(x)$ is an integrable function on the rectangle $T \times T$. By Fubini's theorem

$$(f * g)(t) = \int_{-\pi}^{\pi} f(t - x)g(x)\, dx$$

is well defined for almost all t. We could put an arbitrary value 0 for $f * g$ at the exceptional set where it may not be defined. Now note that

$$2\pi \| f * g \|_1 = \int_{-\pi}^{\pi} |(f * g)(t)|\, dt$$

$$= \int_{-\pi}^{\pi} \left| \int_{-\pi}^{\pi} f(t - x)g(x)\, dx \right| dt \leq \int_{-\pi}^{\pi} \int_{-\pi}^{\pi} |f(t - x)g(x)|\, dx\, dt$$

$$= \int_{-\pi}^{\pi} |g(x)| \left[\int_{-\pi}^{\pi} |f(t - x)|\, dt \right] dx$$

$$= 4\pi^2 \| f \|_1 \| g \|_1 \qquad \blacksquare$$

An immediate application of this is:

Theorem 4.2.2. *Let $f \in L_1(T)$ and let $\sigma_n(f)$ be the Cesàro sums of the Fourier series of f. Then*

$$\lim_{n \to \infty} \| f - \sigma_n(f) \|_1 = 0,$$

i.e., the Fourier series of f is Cesàro summable to f in L_1 norm.

Proof Continuous functions are dense in $L_1(T)$. So given $\varepsilon > 0$ there exists a continuous function g such that $\| f - g \|_1 < \varepsilon$. If F_n denotes the Fejér kernel, defined by (2.8), then by Theorem 2.2.5 and the inequality (4.1) there exists N such that $\| g - g * F_n \|_1 < \varepsilon$ for all $n \geq N$. Hence, we have,

$$\| f - f * F_n \|_1 \leq \| f - g \|_1 + \| g - g * F_n \|_1 + \| g * F_n - f * F_n \|_1$$
$$< 2\varepsilon + \| (g - f) * F_n \|_1.$$

But, by Theorem 4.2.1,

$$\| (g - f) * F_n \|_1 \leq 2\pi \| g - f \|_1 \| F_n \|_1 < 2\pi \varepsilon.$$

So

$$\| f - f * F_n \|_1 < 2(1 + \pi)\varepsilon \quad \text{for all } n \geq N. \qquad \blacksquare$$

Corollary 4.2.3. (The Riemann–Lebesgue Lemma) *If $f \in L_1(T)$, then*

$$\lim_{|n| \to \infty} \hat{f}(n) = 0.$$

Proof Use Theorem 4.2.2 and the idea of the proof of Theorem 2.3.1. $\qquad \blacksquare$

Let c_0 denote the space of sequences converging to 0. The Riemann–Lebesgue Lemma says that the map $f \to \{\hat{f}(n)\}$ is a map from $L_1(T)$ *into* c_0. From results on continuous functions obtained earlier, we know that this map is one-to-one. This map has a particularly pleasing behaviour towards convolutions:

Theorem 4.2.4. *Let f and g be in $L_1(T)$. Then*

$$(\widehat{f * g})(n) = 2\pi \hat{f}(n)\hat{g}(n) \quad \text{for all } n.$$

Proof Once again, using Fubini's Theorem we can write

$$
\begin{aligned}
(\widehat{f * g})(n) &= \frac{1}{2\pi} \int (f * g)(t) e^{-int} \, dt \\
&= \frac{1}{2\pi} \int \int f(t - x) g(x) e^{-int} \, dt \, dx \\
&= \frac{1}{2\pi} \int \int f(t - x) e^{-in(t-x)} g(x) e^{-inx} \, dt \, dx \\
&= \frac{1}{2\pi} \int f(t) e^{-int} \, dt \int g(x) e^{-inx} \, dx \\
&= 2\pi \hat{f}(n) \hat{g}(n).
\end{aligned}
$$

(All integrals above are over the interval $[-\pi, \pi]$.) ∎

Remark. Some books define $f * g$ with a different normalisation as

$$
(f * g)(t) = \frac{1}{2\pi} \int_{-\pi}^{\pi} f(t - x) g(x) \, dx.
$$

With this definition you will get

$$
(\widehat{f * g})(n) = \hat{f}(n) \hat{g}(n)
$$

and

$$
\| f * g \|_1 \le \| f \|_1 \| g \|_1.
$$

A vector space X is called an *algebra* if its elements (x, y, z, etc.) can be multiplied and the product $x \cdot y$ obeys the relations

$$
x \cdot (y \cdot z) = (x \cdot y) \cdot z,
$$

$$
(ax) \cdot y = a(x \cdot y) \quad \text{for all scalars } a.
$$

Suppose X is a Banach space with norm $\|.\|$. If a product on X satisfies the inequality

$$
\| x \cdot y \| \le \| x \| \| y \| \quad \text{for all } x, y,
$$

we say X is a *Banach algebra*.

The space $C(T)$ (with $f + g$ and fg defined as usual) is a Banach algebra with the supremum norm. The space $M(n)$ of $n \times n$ matrices is a Banach algebra with the norm of A defined as $\|A\| = \sup\{\|Au\| : u \in \mathbb{C}^n, \|u\| = 1\}$. The multiplication in $C(T)$ is commutative, that in $M(n)$ is not. Our discussion shows that $L_1(T)$ is a Banach algebra with multiplication of f and g defined by convolution $f * g$ (defined in such a way that $\| f * g \|_1 \le \| f \|_1 \| g \|_1$).

Exercise 4.2.5.

(i) Show that there does not exist any function f in $L_1(T)$ such that $f * g = g$ for all $g \in L_1(T)$.

(ii) Show that there does not exist any continuous function $f \in C(T)$ such that $f * g = g$ for all $g \in C(T)$.

(iii) Let $f \in L_1(T)$ and $g \in C^k(T)$. Show that $f * g$ is in $C^k(T)$, and its kth derivative is $(f * g)^{(k)} = f * g^{(k)}$.

Thus neither the space of continuous functions, nor the space of integrable functions has an identity with respect to convolution. (The constant function 1 is an identity for $C(T)$ with the usual multiplication, the identity matrix is an identity for $M(n)$.)

We have seen that if $f \in L_2(T)$, then $\hat{f}(n)$ not only goes to zero as $|n| \to \infty$, but is square summable. However, for $f \in L_1(T)$ nothing more than $\hat{f}(n) \to 0$ can be said in general. This is proved in Theorem 4.2.10 below.

Exercise 4.2.6. Let f be absolutely continuous and let $f' \in L_2(T)$. Show that

$$\sum_{n=-\infty}^{\infty} |\hat{f}(n)| \le |\hat{f}(0)| + \frac{\pi}{\sqrt{3}} \|f'\|_{L_2}.$$

[Hint: Use the Cauchy–Schwarz Inequality.]

Note that in particular this implies that $\hat{f}(n)$ is in $l_1(Z)$.

Let $\{a_n\}_{n=0}^{\infty}$ be a sequence of real numbers. Let $\Delta a_n = a_{n+1} - a_n$, $\Delta^2 a_n = \Delta(\Delta a_n) = a_{n+2} - 2a_{n+1} + a_n$. Note that $\{a_n\}$ is monotonically increasing if and only if $\Delta a_n \ge 0$ for all n.

Definition 4.2.7. The sequence $\{a_n\}$ is said to be *convex* if $\Delta^2 a_n \ge 0$ for all n.

Proposition 4.2.8. *Suppose the sequence* $\{a_n\}_{n=0}^{\infty}$ *is convex and bounded. Then*

(i) $\{a_n\}$ *is monotonically decreasing and convergent,*

(ii) $\lim n \Delta a_n = 0$,

(iii) $\sum_{n=0}^{\infty} (n+1)\Delta^2 a_n = a_0 - \lim a_n$.

Proof If $\Delta a_n > 0$ for any n, then by the given convexity, $\Delta a_m \ge \Delta a_n > 0$ for all $m \ge n$. But then $\{a_n\}$ cannot be bounded. So $\Delta a_n \le 0$ for all n. This proves (i). Note that

$$\sum_{k=0}^{n} \Delta a_k = a_{n+1} - a_0.$$

Thus the series $\sum(-\Delta a_k)$ is convergent and its terms are positive and decreasing. This proves (ii). We have

$$\sum_{n=0}^{N} (n+1)\Delta^2 a_n = a_0 - a_{N+1} + (N+1)\Delta a_{N+1}.$$

This proves (iii).   ∎

Theorem 4.2.9. *Let* $\{a_n\}_{n=-\infty}^{\infty}$ *be a sequence such that*

(i) $a_n \geq 0$ *for all* n,

(ii) $a_n = a_{-n}$,

(iii) $a_n \to 0$ *as* $n \to \infty$,

(iv) $\{a_n\}_{n=0}^{\infty}$ *is convex.*

Then there exists a nonnegative function $f \in L_1(T)$ *such that* $a_n = \hat{f}(n)$ *for all* n.

Proof Let F_n be the Fejér kernel and

$$f(t) = 2\pi \sum_{n=0}^{\infty} (n+1)\Delta^2 a_n F_{n+1}(t).$$

Since $\|F_n\| = 1/2\pi$ for all n, this series converges in L_1. So $f \in L_1(T)$. Note that $f(t) \geq 0$ since all the terms involved are positive. For each integer k, we have

$$\hat{f}(k) = 2\pi \sum_{n=0}^{\infty} (n+1)\Delta^2 a_n \hat{F}_{n+1}(k).$$

Since

$$F_{n+1}(t) = \frac{1}{2\pi} \sum_{j=-n}^{n} \left(1 - \frac{|j|}{n+1}\right) e^{ijt},$$

we have

$$2\pi \hat{F}_{n+1}(k) = \begin{cases} 1 - |k|/(n+1) & \text{for } |k| \leq n, \\ 0 & \text{for } |k| > n. \end{cases}$$

Hence

$$\hat{f}(k) = \sum_{n=|k|}^{\infty} (n+1)\Delta^2 a_n \left(1 - \frac{|k|}{n+1}\right) = a_{|k|}. \qquad \blacksquare$$

Exercise 4.2.10. There exists a sequence of nonnegative real numbers that is convex and converges to zero more slowly than any given sequence; i.e., if $\{b_n\}$ is any sequence converging to zero, then there exists a convex sequence $\{a_n\}$ such that $|b_n| \leq a_n$ and $a_n \to 0$. [Hint: Think of functions instead of sequences.]

We have shown above that a sequence of nonnegative real numbers can be the sequence of coefficients of a Fourier cosine series and converge to zero at any rate. The situation is different for a sine series:

Theorem 4.2.11. *Let* $f \in L_1(T)$ *and let*

(i) $\hat{f}(n) \geq 0$ *for all* $n \geq 0$,

(ii) $\hat{f}(n) = -\hat{f}(-n)$ *for all* n.

Then

$$\sum_{n \neq 0} \frac{1}{n} \hat{f}(n) < \infty.$$

Proof Let

$$F(x) = \int_{-\pi}^{x} f(t)\, dt, \qquad -\pi \leq x \leq \pi.$$

Then F is absolutely continuous and $F' = f$, a.e. Hence

$$\hat{F}(n) = \frac{1}{in} \hat{f}(n) \quad \text{for all } n \neq 0.$$

Now apply Theorem 2.2.5 to the function F at the point $\theta = 0$ to get

$$F(0) = \hat{F}(0) + \lim_{N \to \infty} 2 \sum_{n=1}^{N} \left(1 - \frac{n}{N+1}\right) \frac{\hat{f}(n)}{in}.$$

So

$$\lim_{N \to \infty} \sum_{n=1}^{N} \left(1 - \frac{n}{N+1}\right) \frac{\hat{f}(n)}{n} = \frac{i}{2}\left(F(0) - \hat{F}(0)\right).$$

Thus the series $\sum \hat{f}(n)/n$ is Cesàro summable; and since its terms are nonnegative it must be convergent. ∎

Corollary 4.2.12. *If a_n are nonnegative real numbers and if*

$$\sum_{n=1}^{\infty} \frac{a_n}{n} = \infty,$$

then the series $\sum_{n=1}^{\infty} a_n \sin nt$ cannot be the Fourier series of any function in $L_1(T)$.

Exercise 4.2.13. Show that the series

$$\sum_{n=2}^{\infty} \frac{\cos nt}{\log n}$$

is the Fourier series of some function in $L_1(T)$, but the series

$$\sum_{n=2}^{\infty} \frac{\sin nt}{\log n}$$

cannot be the Fourier series of any function in $L_1(T)$.

So, what can be said about the class $\mathcal{A}$ of all sequences which are Fourier coefficients of functions in $L_1(T)$? We have shown the following:

(i) $\mathcal{A} \subset c_0$, the space of doubly infinite sequences going to zero at $\pm\infty$,

(ii) $\mathcal{A} \neq c_0$,

(iii) $\mathcal{A}$ is an algebra, i.e., it is closed under addition and (pointwise) multiplication because of the relations:

$$(\widehat{f + g})(n) = \hat{f}(n) + \hat{g}(n),$$
$$(\widehat{f * g})(n) = 2\pi \hat{f}(n)\hat{g}(n).$$

However, a complete description of the class $\mathcal{A}$ is not known. (In fact, it is believed that there is *no* good description of $\mathcal{A}$).

5

Some Applications

One test of the depth of a mathematical theory is the variety of its applications. We began this study with a problem of heat conduction. The solution of this problem by Fourier series gives rise to an elegant mathematical theory with several applications in diverse areas. We have already seen some of them. We have proved the Weierstrass approximation theorem in Chapters 1 and 3. This is a fundamental theorem useful in numerical analysis, in approximation theory and in other branches of analysis. In Chapter 3 and 4 the Dirichlet and Plancherel theorems were used to obtain the sums of some series and the values of some integrals. Some more applications of Fourier series are described below.

5.1 An ergodic theorem and number theory

The *ergodic principle* in statistical mechanics states the following: *the time average of a mechanical quantity should be the same as its phase average.* Rather than explain the general principle we will find it instructive to see how it operates in a special situation. The model we describe is due to H. Weyl.

Consider the circle $T = [-\pi, \pi)$. Let $\varphi \in (-\pi, \pi)$ and define a map $R_\varphi : T \to T$ by

$$R_\varphi(\theta) = \theta + \varphi \bmod 2\pi. \tag{5.1}$$

If T is thought of as the unit circle in the plane, then R_φ is a rotation by angle φ. We think of T as *the phase space* and the action of R_φ as *dynamics* on this space. If $\theta = \theta_0$ is a given point on T, then its *trajectory* under this dynamics is the succession of points

$$
\begin{aligned}
\theta_0 &= \theta, \\
\theta_1 &= R_\varphi\theta = \theta + \varphi \quad \bmod 2\pi, \\
\theta_2 &= R_\varphi^2\theta = \theta + 2\varphi \quad \bmod 2\pi, \\
&\vdots \\
\theta_k &= R_\varphi^k\theta = \theta + k\varphi \quad \bmod 2\pi.
\end{aligned}
$$

Any continuous function f on T is thought of as a *mechanical quantity*. The *time average* of f is defined as

$$\lim_{N \to \infty} \frac{1}{N} \sum_{k=0}^{N-1} f(\theta_k), \tag{5.2}$$

and its *phase average* is defined as

$$\frac{1}{2\pi} \int_{-\pi}^{\pi} f(\theta) \, d\theta. \tag{5.3}$$

One can think of θ_0 moving to θ_1, θ_2, ... at successive time intervals. The sum in (5.2) then gives the average value of f on the trajectory of f from time 0 to N. The integral (5.3), on the other hand, gives the average value of f over the phase space T.

An example of the ergodic principle is:

Theorem 5.1.1. (Weyl) *If φ is an irrational multiple of 2π, then the quantities (5.2) and (5.3) are equal.*

Proof Given a continuous function f write $W_N(f)$ for the difference between (5.2) and (5.3):

$$W_N(f) = \frac{1}{N} \sum_{k=0}^{N-1} f(\theta_k) - \frac{1}{2\pi} \int_{-\pi}^{\pi} f(\theta) \, d\theta.$$

We want to show $W_N(f) \to 0$ as $N \to \infty$. The idea of the proof is to do this first for the functions $e_n(\theta) = e^{in\theta}$, then for trigonometric polynomials and then by Fejér's Theorem for any continuous function f.

When $n = 0$, $e_n(\theta) \equiv 1$; and for $f \equiv 1$ both (5.2) and (5.3) are equal to 1. So, in this case $W_N = 0$ for all N.

When $n \neq 0$ we can write

$$|W_N(e_n)| = \left| \frac{1}{N} \sum_{k=0}^{N-1} e^{in(\theta + k\varphi)} - \frac{1}{2\pi} \int_{-\pi}^{\pi} e^{in\theta} \, d\theta \right|$$

$$= \left| \frac{e^{in\theta}}{N} \sum_{k=0}^{N-1} e^{ink\varphi} \right|$$

$$= \left| \frac{e^{in\theta}}{N} \frac{1 - e^{inN\varphi}}{1 - e^{in\varphi}} \right|$$

$$\leq \frac{2}{N} \left| \frac{1}{1 - e^{in\varphi}} \right|.$$

Note that $e^{in\varphi} \neq 1$ since φ is an irrational multiple of 2π. As $N \to \infty$, the right-hand side of the inequality above goes to 0.

Now note that both (5.2) and (5.3) are linear functions of f. So

$$W_N(p) \to 0 \quad \text{as} \quad N \to \infty,$$

whenever p is an exponential polynomial.

This linearity also shows that if f and g are continuous functions on T such that

$$\sup |f(\theta) - g(\theta)| < \varepsilon,$$

then

$$|W_N(f) - W_N(g)| < 2\varepsilon, \quad \text{for all } N.$$

Now use Fejér's Theorem to complete the proof. ∎

Exercise 5.1.2. Show that if φ is a rational multiple of 2π, then (5.2) and (5.3) are not always equal.

A remarkable consequence of Theorem 5.1.1 is *Weyl's Equidistribution Theorem* in number theory. Let $\tilde{x}$ denote the fractional part of a real number x, i.e., $0 \le \tilde{x} < 1$ and $x - \tilde{x}$ is an integer. Weyl's equidistribution theorem says that if x is irrational, then for large N the fractional parts of $x, 2x, \ldots, Nx$ are uniformly distributed over $(0, 1)$. A precise statement is:

Theorem 5.1.3. (Weyl) *If x is an irrational number, then for every subinterval $[a, b]$ of $(0, 1)$,*

$$\lim_{N \to \infty} \frac{1}{N} \operatorname{card}\{k : 1 \le k \le N, (kx)^{\sim} \in [a, b]\} = b - a. \tag{5.4}$$

Here card A *denotes the cardinality of a set A, and $(kx)^{\sim}$ is the fractional part of the number kx.*

Proof By a simple change of variables the statement (5.4) is reduced to an equivalent statement: if $x \in T$ and is an irrational multiple of 2π, then for every subinterval $[a, b]$ of $(-\pi, \pi)$,

$$\lim_{N \to \infty} \frac{1}{N} \operatorname{card}\{k : 1 \le k \le N, (kx)^{\sim} \in [a, b]\} = \frac{b - a}{2\pi}. \tag{5.5}$$

Of course, here kx is again thought of as an element of T for all k.

We will prove (5.5) by applying Theorem 5.1.1. Let $\chi_{[a,b]}$ denote the characteristic function of $[a, b]$, i.e.,

$$\chi_{[a,b]}(t) = \begin{cases} 1 & \text{if } t \in [a, b], \\ 0 & \text{otherwise.} \end{cases}$$

This is a discontinuous function. Given $\varepsilon > 0$ we can choose two functions f_+ and f_- which are continuous and approximate $\chi_{[a,b]}$ from inside and outside to within ε, i.e.,

$$0 \leq f_-(t) \leq \chi_{[a,b]}(t) \leq f_+(t) \leq 1 \quad \text{for all } t \tag{5.6}$$

and

$$(b-a) - \varepsilon \leq \int_{-\pi}^{\pi} f_-(t)\, dt \leq \int_{-\pi}^{\pi} f_+(t)\, dt \leq (b-a) + \varepsilon. \tag{5.7}$$

Note that (5.6) implies

$$\sum_{k=1}^{N} f_- \left((kx)^{\sim}\right) \leq \operatorname{card}\{k : 1 \leq k \leq N, (kx)^{\sim} \in [a,b]\} \leq \sum_{k=1}^{N} f_+ \left((kx)^{\sim}\right). \tag{5.8}$$

By Theorem 5.1.1, there exists N_0 such that for all $N \geq N_0$,

$$\left| \sum_{k=1}^{N} f_{\pm} \left((kx)^{\sim}\right) - \frac{1}{2\pi} \int_{-\pi}^{\pi} f_{\pm}(t)\, dt \right| \leq \varepsilon. \tag{5.9}$$

Now use (5.7), (5.8) and (5.9) to obtain (5.5). ∎

Notice that we have interpreted the process of picking up the fractional parts of $x, 2x, 3x, \ldots$ as a "dynamical process."

Exercise 5.1.4. Show that the statement of Theorem 5.1.3 is not true when x is a rational number.

5.2 The isoperimetric problem

Among all simple closed plane curves with a given perimeter, which one encloses the maximum area? (Here a "simple" curve means a curve that does not intersect itself.) This is called the isoperimetric problem. The Greek founders of geometry proved that a circle of perimeter L always encloses a larger area than any *polygon* with the same perimeter. This result is attributed to Zenodorus (famous for "Zeno's paradox"). In 1841 Steiner proved that among all simple closed plane curves with perimeter L, the circle encloses the maximum area. A very elegant proof of this fact using Fourier series was given by Hurwitz. We will give this proof here under the additional assumption that all curves that we consider are (piecewise) C^1.

Let C be any curve in the plane parametrised over $[-\pi, \pi]$, i.e., C is the range of a continuous map $(x(t), y(t))$ from $[-\pi, \pi]$ to $\mathbb{R}^2$. You can think of C as the trajectory of a particle moving in the plane from time $-\pi$ to π. We assume that the curve is

(i) closed, i.e., $(x(-\pi), y(-\pi)) = (x(\pi), y(\pi))$,
(ii) simple, i.e., if $-\pi \leq s < t < \pi$ then $(x(s), y(s)) \neq (x(t), y(t))$,
(iii) smooth, i.e., $x(t)$ and $y(t)$ both are C^1 functions,

(iv) of length L, i.e.,

$$\int_{-\pi}^{\pi} [(x'(t))^2 + y'(t))^2]^{1/2} \, dt = L.$$

It is convenient to choose the parametrisation in such a way that

$$(x'(t))^2 + (y'(t))^2 = \frac{L^2}{4\pi^2}. \tag{5.10}$$

In the picture of the moving particle this means that the particle moves with uniform speed while tracing the curve C. This involves no loss of generality in our problem.

If A is the area enclosed by such a curve, then

$$A = \int_{-\pi}^{\pi} x(t) y'(t) \, dt. \tag{5.11}$$

Integrate (5.10) over $[-\pi, \pi]$ and use Plancherel's Theorem (Exercise 4.1.10). This gives

$$\frac{L^2}{4\pi^2} = \frac{1}{2\pi} \int_{-\pi}^{\pi} [(x'(t))^2 + (y'(t))^2] \, dt$$

$$= \sum_{n=-\infty}^{\infty} (|\hat{x}'(n)|^2 + |\hat{y}'(n)|^2)$$

$$= \sum_{n=-\infty}^{\infty} n^2 (|\hat{x}(n)|^2 + |\hat{y}(n)|^2) \tag{5.12}$$

by Exercise 2.3.11.

Now note that, since $x(t)$ and $y(t)$ both are real functions, $\hat{x}(n)$ and $\hat{y}(n)$ are complex conjugates of $\hat{x}(-n)$ and $\hat{y}(-n)$ respectively. So (5.12) can be written as

$$L^2 = 8\pi^2 \sum_{n=1}^{\infty} n^2 (|\hat{x}(n)|^2 + |\hat{y}(n)|^2). \tag{5.13}$$

Similarly, using Parseval's relations (Exercise 4.1.13) we get from (5.11)

$$A = 2\pi \sum_{n=-\infty}^{\infty} \hat{x}(n) \overline{(\hat{y}'(n))}$$

$$= -2\pi i \sum_{n=-\infty}^{\infty} n \hat{x}(n) \overline{\hat{y}(n)}$$

$$= 4\pi \sum_{n=1}^{\infty} n \, \text{Im}(\hat{x}(n) \overline{\hat{y}(n)}). \tag{5.14}$$

Here Im denotes the imaginary part and the bar denotes complex conjugation.

From (5.13) and (5.14) we get

$$L^2 - 4\pi A = 8\pi^2 \sum_{n=1}^{\infty} \{n^2 (|\hat{x}(n)|^2 + |\hat{y}(n)|^2) - 2n \, \text{Im} \, \hat{x}(n) \overline{\hat{y}(n)}\}. \tag{5.15}$$

Now if $\hat{x}(n) = \alpha_n + i\beta_n$ and $\hat{y}(n) = \gamma_n + i\delta_n$ are the respective real-imaginary decompositions, then the nth term of (5.15) can be written as

$$n^2(\alpha_n^2 + \beta_n^2 + \gamma_n^2 + \delta_n^2) - 2n(\beta_n\gamma_n - \alpha_n\delta_n)$$
$$= (n^2 - 1)(\alpha_n^2 + \beta_n^2) + (\alpha_n + n\delta_n)^2 + (\beta_n - n\gamma_n)^2,$$

and this is nonnegative for each n. Hence from (5.15) we get

$$L^2 \geq 4\pi A$$

and equality holds here if and only if

$$\alpha_n = \beta_n = \gamma_n = \delta_n = 0 \quad \text{for } n \geq 2,$$
$$\alpha_1 = -\delta_1, \quad \beta_1 = \gamma_1.$$

Hence, we have $L^2 = 4\pi A$ if and only if

$$x(t) = \hat{x}(0) + 2(\alpha_1 \cos t - \beta_1 \sin t),$$
$$y(t) = \hat{y}(0) + 2(\beta_1 \cos t + \alpha_1 \sin t),$$

or equivalently, if and only if

$$(x(t) - \hat{x}(0))^2 + (y(t) - \hat{y}(0))^2 = 4(\alpha_1^2 + \beta_1^2) \quad \text{for all } t.$$

This means that the curve C is a circle.

We have proved that among all smooth simple closed curves in the plane with a given perimeter, the circle encloses the maximum area. With a little work the condition of smoothness can be dropped. You can try proving this by assuming the following: if C is a rectifiable simple closed curve in the plane with perimeter L and enclosed area A, then there exists a sequence of smooth simple closed curves C_n such that C_n converges uniformly to C; further if L_n and A_n are, respectively, the perimeter of C_n and the area enclosed by C_n, then $L_n \to L$ and $A_n \to A$.

The isoperimetric problem is believed to be the first *extremal problem* discussed in the scientific literature. It is also known as *Dido's Problem*. The epic poem *Aeneid* by the poet Vergil of ancient Rome tells a legend: *Fleeing from persecution by her brother, the Phoenician princess Dido set off westward along the Mediterranean shore in search of a haven. A certain spot on the coast of what is now the bay of Tunis caught her fancy. Dido negotiated the sale of land with the local leader, Yarb. She asked for very little—as much as could be "encircled with a bull's hide." Dido managed to persuade Yarb, and a deal was struck. Dido then cut a bull's hide into narrow strips, tied them together, and enclosed a large tract of land. On this land she built a fortress and, near it, the city of Carthage. There she was fated to experience unrequited love and a martyr's death.*

This quote is from *Stories about Maxima and Minima* by V.M. Tikhomirov. We recommend this book, and Chapter VII of the classic *What Is Mathematics?* by R. Courant and H. Robbins, where several extremal problems arising in mathematics, physics, astronomy, and other areas are discussed.

5.3 The vibrating string

The solution by d'Alembert of the problem of the vibrating string was the beginning of the development of Fourier series. The connection is explained in this section.

Just as we derived the Laplace equation for steady state heat conduction from a simple law—the Newton law of cooling—we can derive the equation governing the vibrations of a string using another simple law of Newton—the Second Law of Motion.

Imagine a thin, long, stretched elastic string. If the length is much greater than the thickness we may think of it as a one-dimensional string stretched between points 0 and L on the x-axis. If it is plucked in the y direction and released, it will start vibrating. We assume that the tension of the string is high and the vibrations are small, all motion takes place in the x-y plane and all points of the string move perpendicular to the x-axis. These assumptions, no doubt, are simplistic but they describe the behaviour of a typical system— such as a musical instrument—fairly accurately.

Let $u(x, t)$ describe the profile of the string, i.e., $u(x, t)$ is the displacement in the y-direction at time t of a point which was originally at point x in the equilibrium position.

Consider a small portion of the string between the points a and b, $a < b$. If ρ is the density of the string, the mass of this portion is $\rho(b - a)$ and its acceleration is approximately $\partial^2 u / \partial t^2$ evaluated at some point x, $a \leq x \leq b$. So, by Newton's second law of motion the force acting on this portion of the string is

$$f = \rho(b - a)\frac{\partial^2 u}{\partial t^2}. \tag{5.16}$$

We can also calculate this force by another argument. This force arises from the tension τ of the string. We can assume τ is constant when vibrations are small. This force acts along the string. Its component in the y-direction is obtained by multiplying it by the sine of the angle of inclination, and this is $[1 + (\partial u/\partial x)^2]^{-1/2}\partial u/\partial x$. Therefore, if $\partial u/\partial x$ is much smaller than 1, as is the case when the vibrations are small, this is nearly equal to $\partial u/\partial x$. So the total force acting on the portion of the string between a, b is approximately

$$f = \tau \frac{\partial u}{\partial x}\Big|_a^b. \tag{5.17}$$

Now equate (5.16) and (5.17), divide by $b - a$ and let $b \to a$. This gives

$$\frac{\partial^2 u}{\partial t^2} = \frac{\tau}{\rho}\frac{\partial^2 u}{\partial x^2}.$$

This is usually written as

$$\frac{\partial^2 u}{\partial t^2} = a^2\frac{\partial^2 u}{\partial x^2}, \tag{5.18}$$

and called the *one-dimensional wave equation*.

Remark 5.3.1. Note that by a "dimension analysis" we can see what the factor a in (5.18) represents. The units of τ are those of force, i.e., mass $\times$ length $\times$ (time)$^{-2}$, the units of ρ

are those of density, i.e., mass $\times$ (length)$^{-1}$. So τ/ρ has the units of (length)^2 $\times$ (time)$^{-2}$, i.e., of (velocity)^2. Thus a is a quantity like a velocity. This can be seen by analysing (5.18) also. We will find this information useful, even though it does not affect any mathematical calculation. In fact, in several mathematics books you will find the statement "a is a constant, which may be put equal to 1 without loss of generality."

In addition to (5.18) our system satisfies the *boundary conditions*

$$u(0, t) = u(L, t) = 0 \quad \text{for all } t. \tag{5.19}$$

This says that the end points of the string are fixed throughout the motion. The initial position and velocity can be chosen at will. This is stated as *initial conditions*

$$u(x, 0) = f(x), \qquad u_t(x, 0) = g(x), \tag{5.20}$$

where f and g describe the initial position and the initial velocity of a particle on the string at the position x. Since we are assuming that u satisfies (5.18), both f and g must be C^2 functions (at least piecewise C^2).

The problem is to find u that satisfies (5.18) – (5.20).

One way is to proceed as in Section 1.2 where we studied the heat problem. Assume that there is a solution of the form

$$u(x, t) = X(x)T(t), \tag{5.21}$$

where X is a function of x only and T is a function t only. Then (5.18) leads to the equation

$$\frac{X''}{X} = \frac{1}{a^2}\frac{T''}{T}.$$

Since the left-hand side is independent of t, and the right-hand side is independent of x, there is a constant c such that

$$\frac{X''}{X} = \frac{1}{a^2}\frac{T''}{T} = -c.$$

This leads to two second order *ordinary* differential equations

$$X'' + cX = 0,$$

$$T'' + a^2 cT = 0. \tag{5.22}$$

Both equations are of the type (1.6) encountered earlier.

When $c = 0$, the equations (5.22) lead to the solution

$$u(x, t) = (Ax + B)(Ct + D),$$

where A, B, C, D are constants. Now if $C \neq 0$, $u(x, t)$ becomes unbounded as $t \to \infty$. We know from physical considerations that u must remain bounded. So $C = 0$. This means that u is independent of t. So there cannot be any motion. This is contrary to what we know.

If $c \neq 0$, then proceeding as in Section 1.2 we get the solution

$$u(x, t) = (Ae^{ix\alpha} + Be^{-ix\alpha})(Ce^{it\beta} + De^{-it\beta}), \tag{5.23}$$

where $\alpha = \sqrt{c}, \beta = \sqrt{c}a$ and A, B, C, D are constants. If $c > 0$, then both α, β are real and we can assume without loss of generality that both are positive—otherwise we can interchange A and B, and C and D. If $c < 0$, then both α, β are imaginary.

From the first condition in (5.19), and (5.23) we get

$$0 = u(0, t) = (A + B)(Ce^{it\beta} + De^{-it\beta})$$

for all t. If the second factor on the right-hand side were zero, then $u(x, t)$ would be identically zero. This is not possible. Hence

$$A = -B. \tag{5.24}$$

Now using the second of the boundary conditions (5.19), (5.23), and (5.24) we get

$$A(e^{iL\alpha} - e^{iL\alpha}) = 0. \tag{5.25}$$

If $c < 0$, then $\alpha = i\gamma$ for some real $\gamma \neq 0$, and (5.25) becomes

$$A(e^{-L\gamma} - e^{L\gamma}) = 0,$$

which implies $A = 0$ and hence from (5.24) $B = 0$ as well. Again, this would mean that $u(x, t)$ is identically zero, and that is not possible.

So, we are left with the case $c > 0$. Now α is real and positive, and (5.25) gives

$$A \sin L\alpha = 0.$$

Since A cannot be zero (as we saw above) this condition implies that

$$\alpha = \frac{n\pi}{L} \qquad n = 1, 2, \ldots.$$

Then we must have

$$\beta = \frac{n\pi}{L}a.$$

Hence for each positive integer n we have a solution of (5.18) given by

$$u_n(x, t) = \sin \frac{n\pi x}{L}(A_n e^{intw} + B_n e^{-intw}), \tag{5.26}$$

where A_n, B_n are some (new) constants and $w = \pi a/L$.

A finite linear sum of such u_n would be a solution of (5.20) and so would be the infinite sum

$$u(x, t) = \sum_{n=1}^{\infty} \sin \frac{n\pi x}{L}(A_n e^{intw} + B_n e^{-intw}), \tag{5.27}$$

provided this series is convergent and can be differentiated term by term.

Exercise 5.3.2. Show that the assumption $u \in C^2$ implies that the series (5.27) converges uniformly and can be differentiated term by term.

The arbitrary constants A_n and B_n occurring in (5.27) are determined from the initial condition (5.20). These give

$$f(x) = \sum_{n=1}^{\infty} (A_n + B_n) \sin \frac{n\pi x}{L},$$

$$g(x) = \sum_{n=1}^{\infty} inw(A_n - B_n) \sin \frac{n\pi x}{L}. \tag{5.28}$$

Thus $A_n + B_n$ and $inw(A_n - B_n)$ are the coefficients in the Fourier sine series for f and g respectively. So, they are determined uniquely by f and g. Hence A_n and B_n are uniquely determined by f and g.

Remark 5.3.3. Since the quantity a is a velocity, as observed earlier, the quantity w has the units of velocity $\times$ (length)$^{-1}$ = (time)$^{-1}$. Thus w can be thought of as a *frequency*. Our solution (5.27) involves positive integral multiples nw of this frequency. In music w is called the *fundamental* and $2w, 3w, \ldots$ are called the *overtones*. Note that

$$w = \frac{\pi}{L} a = \frac{\pi}{L} \sqrt{\frac{\tau}{\rho}}.$$

So w increases with tension and decreases with the length and the density of the string. This corresponds well with our practical experience.

D'Alembert's original solution was different from (5.27). We can get it from (5.27) as follows. Expand the exponentials occuring there into cosines and sines, then use familiar trigonometric identities to get (5.27) into the form

$$u(x, t) = \frac{1}{2} \sum (A_n + B_n) \left[\sin \left(\frac{n\pi x}{L} + ntw \right) + \sin \left(\frac{n\pi x}{L} - ntw \right) \right]$$

$$- \frac{1}{2} \sum i (A_n - B_n) \left[\cos \left(\frac{n\pi x}{L} + ntw \right) - \cos \left(\frac{n\pi x}{L} - ntw \right) \right].$$

Now, replace w by $\pi L / a$ and rearrange terms to get

$$u(x, t) = \frac{1}{2} \sum \left[(A_n + B_n) \sin \frac{n\pi}{L} (at + x) - i(A_n - B_n) \cos \frac{n\pi}{L} (at + x) \right]$$

$$- \frac{1}{2} \sum \left[(A_n + B_n) \sin \frac{n\pi}{L} (at - x) - i(A_n - B_n) \cos \frac{n\pi}{L} (at - x) \right]. \tag{5.29}$$

So, one can write

$$u(x, t) = v(at + x) - v(at - x) \tag{5.30}$$

where v is defined by (5.29) and (5.30). Note that v is a function of *one* variable.

D'Alembert's solution was given in the form (5.30). It is not necessary to come to it via the route followed above. A much simpler way is the following. Let v be any C^2 function of period $2L$. Then if we define $u(x, t)$ by (5.30), we can easily check that it satisfies the wave equation (5.18), and also the boundary conditions (5.19). If $u(x, t)$ is now to satisfy the initial conditions (5.20), then we must have

$$f(x) = u(x, 0) = v(x) - v(-x),$$

$$g(x) = u_t(x, 0) = v'(x) - v'(-x).$$

These two conditions completely determine v; the first one determines the odd part of v and the second determines its even part.

Notice that (5.30) can be interpreted as a superposition of two waves travelling in the opposite direction with the same speed a. As a function of x, the graph of $v(at + x)$ has the same shape as that of $v(x)$ but is translated by a distance at to the left. This can be thought of as a wave travelling to the left with speed a. The point 0 on the string is fixed. The function $-v(at - x)$ represents the wave $v(at + x)$ reflected at the origin, and now travelling to the right with speed a. The solution (5.30) is a superposition of these two waves.

5.4 Band matrices

In this section we demonstrate an application of ideas from Fourier series to a problem in linear algebra.

Let $M(n)$ be the space of $n \times n$ complex matrices. A matrix A is also a linear operator on the Euclidean space $\mathbb{C}^n$. The *norm* of A is its norm as an operator, defined as

$$\|A\| = \sup\{\|Ax\| : x \in \mathbb{C}^n, \|x\| = 1\}. \tag{5.31}$$

An operator U is said to be *unitary* if it satisfies any of the following equivalent conditions:

1. U preserves inner products, i.e., $\langle Ux, Uy \rangle = \langle x, y \rangle$ for all $x, y, \in \mathbb{C}^n$.
2. U preserves norms, i.e., $\|Ux\| = \|x\|$ for all $x \in \mathbb{C}^n$.
3. U is invertible, and (its Hermitian conjugate) $U^* = U^{-1}$; i.e., $UU^* = U^*U = I$, the identity operator.
4. For each orthonormal basis $e_1, \ldots, e_n$ in $\mathbb{C}^n$, the vectors $Ue_1, \ldots, Ue_n$ also form an orthonormal basis.

It is easy to see that

$$\|UAV\| = \|A\| \quad \text{for all unitary } U, V. \tag{5.32}$$

Exercise 5.4.1. Let $\omega = e^{2\pi i/n}$, a primitive nth root of unity. For $0 \le j \le n - 1$, let f_j be the n-vector with components

$$f_{jk} = \frac{1}{\sqrt{n}}\omega^{jk}, \qquad 0 \le k \le n - 1.$$

Thus

$$f_0 = \frac{1}{\sqrt{n}}(1, 1, \ldots, 1),$$

$$f_1 = \frac{1}{\sqrt{n}}(1, \omega, \omega^2, \ldots, \omega^{n-1}),$$

$$\vdots$$

etc. Show that $f_0, f_1, \ldots, f_{n-1}$ is an orthonormal basis for $\mathbb{C}^n$. Show that the matrix F whose jth column is the vector f_j is a unitary matrix.

Exercise 5.4.2. The permutation matrix

$$R = \begin{bmatrix} 0 & 0 & \cdots & 1 \\ 1 & 0 & \cdots & 0 \\ 0 & 1 & \cdots & 0 \\ \cdot & \cdot & \cdot & \cdot \\ 0 & 0 & 1 & 0 \end{bmatrix}$$

is unitary. Let F be the matrix of Exercise 5.4.1 and let $U = FRF^*$. Show that U is the diagonal unitary matrix with diagonal entries $1, \omega, \ldots, \omega^{n-1}$, i.e.,

$$U = \text{diag}(1, \omega, \omega^2, \ldots, \omega^{n-1}). \tag{5.33}$$

Exercise 5.4.3. A few important facts of Linear Algebra are listed in this exercise. Prove them.

(i) The Spectral Theorem says that every Hermitian operator (one for which $A = A^*$) can be diagonalised by a "unitary conjugation"; i.e., there exists a unitary U such that

$$UAU^* = \text{diag}(\lambda_1, \ldots, \lambda_n).$$

The numbers λ_j are the eigenvalues of A. So, for a Hermitian operator,

$$\|A\| = \max_{1 \le j \le n} |\lambda_j|.$$

(ii) For any operator A we have $\|A\|^2 = \|A^*A\|$.

(iii) The positive square roots of the eigenvalues of A are called the singular values of A. They are enumerated in decreasing order as $s_1(A) \ge s_2(A) \ge \cdots \ge s_n(A)$. We have $\|A\| = s_1(A)$.

Exercise 5.4.4. Calculate the norms of the 2×2 matrices

$$A = \begin{bmatrix} 1 & 1 \\ -1 & 1 \end{bmatrix}, \qquad B = \begin{bmatrix} 1 & 1 \\ 0 & 1 \end{bmatrix}.$$

You will find that $\|A\| = \sqrt{2}$, and $\|B\| = \frac{1}{2}(1 + \sqrt{5})$. In this example, B is obtained by replacing one of the entries of A by zero. But $\|B\| > \|A\|$.

The rest of this section is concerned with finding relationships between $\|B\|$ and $\|A\|$ when B is obtained from A by replacing some of its entries by zeros.

Divide A into r^2 blocks in which the diagonal blocks are square matrices, not necessarily of the same sizes. The block-diagonal matrix obtained by replacing the off-diagonal blocks $A_{ij}, i \neq j$, by zeros is called a *pinching* (an *r-pinching*) of A. We write this as

$$\mathcal{C}(A) = \operatorname{diag}(A_{11}, \ldots, A_{rr}). \tag{5.34}$$

Thus a 2-pinching is obtained by splitting A as $[A_{ij}], i, j = 1, 2$, and then putting

$$\mathcal{C}(A) = \begin{bmatrix} A_{11} & 0 \\ 0 & A_{22} \end{bmatrix}. \tag{5.35}$$

Exercise 5.4.5.

(i) Let U be the block-diagonal matrix $U = \operatorname{diag}(I, -I)$ where the blocks have the same sizes as in the decomposition (5.35). Show that the 2-pinching

$$\mathcal{C}(A) = \frac{1}{2}(A + UAU^*). \tag{5.36}$$

Hence $\mathcal{C}(A) \leq \|A\|$.

(ii) Show that the r-pinching (5.34) can be obtained from A by successively applying $(r - 1)$ 2-pinchings. Thus $\mathcal{C}(A) \leq \|A\|$ for every pinching.

(iii) Show that there exist r mutually orthogonal projections P_j in $\mathbb{C}^n$, $P_1 + \cdots + P_r = I$, such that the pinching (5.34) can be written as

$$\mathcal{C}(A) = \sum_{j=1}^{r} P_j A P_j. \tag{5.37}$$

When $r = n$, the pinching (5.34) reduces to a diagonal matrix. We write this as $\mathcal{D}(A)$ and call it the *diagonal part* of A. In this case the projections P_j in (5.37) are the projections onto the 1-dimensional spaces spanned by the basis vectors.

Exercise 5.4.6.

(i) Let U be the diagonal unitary matrix (5.33). Show that

$$\mathcal{D}(A) = \frac{1}{n} \sum_{j=0}^{n-1} U^j A U^{\star j}. \tag{5.38}$$

(ii) Any r-pinching can be described as

$$\mathcal{C}(A) = \frac{1}{r} \sum_{j=0}^{r-1} V^j A V^{\star j} \tag{5.39}$$

where V is a diagonal unitary matrix. What is V?

(iii) Use this representation to give another proof of the inequality $\|\mathcal{C}(A)\| \le \|A\|$.

The expressions (5.37) and (5.39) display a pinching as a (*noncommutative*) *averaging operation* or a *convex combination* of unitary conjugates of A.

For $1 \le j \le n-1$, let $\mathcal{D}_j(A)$ be the matrix obtained from A by replacing all its entries except those on the jth superdiagonal by zeros. (A superdiagonal is a diagonal parallel to the main diagonal and above it.) Likewise, let $\mathcal{D}_{-j}(A)$ be the matrix obtained by retaining only the jth subdiagonal of A. In consonance with this notation we say $\mathcal{D}_0(A) = \mathcal{D}(A)$.

Let U_θ be the diagonal matrix

$$U_\theta = \operatorname{diag}(e^{i\theta}, e^{2i\theta}, \ldots, e^{ni\theta}).$$

The (r, s) entry of the matrix $U_\theta A U_\theta^\star$ is then $e^{i(r-s)\theta} a_{rs}$. Hence

$$\mathcal{D}_k(A) = \frac{1}{2\pi} \int_{-\pi}^{\pi} e^{ik\theta} U_\theta A U_\theta^\star \, d\theta. \tag{5.40}$$

When $k = 0$, this gives one more representation of the main diagonal $\mathcal{D}(A)$ as an average over unitary conjugates of A. From this we get the inequality $\|\mathcal{D}_k(A)\| \le \|A\|$. Using (5.40) we see that

$$\|\mathcal{D}_k(A) + \mathcal{D}_{-k}(A)\| \le \frac{1}{2\pi} \int_{-\pi}^{\pi} |2\cos k\theta| \, d\theta \|A\|.$$

Let $\mathcal{T}_3(A) = \mathcal{D}_{-1}(A) + \mathcal{D}_0(A) + \mathcal{D}_1(A)$. This is the matrix obtained from A by keeping its middle three diagonals and replacing the other entries by zeros. Again, using (5.40) we see that

$$\|\mathcal{T}_3(A)\| \le \frac{1}{2\pi} \int_{-\pi}^{\pi} |1 + 2\cos\theta| \, d\theta \|A\|.$$

Exercise 5.4.7. Calculate the last two integrals. You will get

$$\|\mathcal{D}_k(A) + \mathcal{D}_{-k}(A)\| \le \frac{4}{\pi} \|A\|, \tag{5.41}$$

$$\|\mathcal{T}_3(A)\| \le \left(\frac{1}{3} + \frac{2\sqrt{3}}{\pi} \right) \|A\|. \tag{5.42}$$

Note that using the inequality $\|\mathcal{D}_k(A)\| \le \|A\|$ and the triangle inequality we get the bounds $2\|A\|$ and $3\|A\|$, respectively, for the left-hand sides of (5.41) and (5.42). These bounds are weaker than what we have obtained from the integrals.

A *trimming* of A is a matrix of the form

$$\mathcal{T}_{2k+1}(A) = \sum_{j=-k}^{k} \mathcal{D}_j(A),$$

obtained by replacing all diagonals of A outside the band $-k \le j \le k$ by zeros. We have

$$\mathcal{T}_{2k+1}(A) = \int_{-\pi}^{\pi} D_k(\theta) U_\theta A U_\theta^\star \, d\theta,$$

where $D_k(\theta)$ is the Dirichlet kernel (2.6). Hence, we have

$$\|\mathcal{T}_{2k+1}(A)\| \le L_k \|A\|, \tag{5.43}$$

where L_k is the Lebesgue constant. Since $L_k = O(\log k)$ this inequality is much stronger than the inequality $\|\mathcal{T}_{2k+1}(A)\| \le (2k+1)\|A\|$ that one gets by using just the triangle inequality and $\|\mathcal{D}_j(A)\| \le \|A\|$.

Let $\Delta_U(A)$ be the matrix obtained from A by replacing all entries of A below the main diagonal by zeros. Then Δ_U is called the *triangular truncation operator.*

Exercise 5.4.8.

(i) Let B be a $k \times k$ matrix and let

$$A = \begin{bmatrix} O & B^\star \\ B & O \end{bmatrix}.$$

This is a matrix of order $2k$. Show that $\|A\| = \|B\|$.

(ii) Show that

$$\mathcal{T}_{2(k+1)+1}(A) = \begin{bmatrix} O & \Delta_U(B)^\star \\ \Delta_U(B) & O \end{bmatrix}.$$

(iii) Thus, if Δ_U is the triangular truncation operator on $M(k)$, then

$$\|\Delta_U(B)\| \le L_{k+1} \|B\|, \tag{5.44}$$

where L_{k+1} is the Lebesgue constant.

Let A be the $k \times k$ matrix with entries $a_{ij} = (i - j)^{-1}$ if $i \ne j$, and $a_{ii} = 0$. It can be seen (with some labour) that $\|A\| \le \pi$, while $\|\Delta_U(A)\| = O(\log k)$. Thus the inequality (5.44) is nearly the best possible. This is an important fact. It corresponds to a basic fact about Fourier series proved by M. Riesz. The *analytic part* of a Fourier series $\sum_{n=-\infty}^{\infty} a_n e^{in\theta}$ is the truncated series $\sum_{n=0}^{\infty} a_n e^{in\theta}$. Riesz showed that the truncation map that sends the Fourier series of a function to its analytic part is a bounded map on the space L_p if $1 < p < \infty$, but not if $p = 1$ or ∞.

The matrix F of Exercise 5.4.1 is called the matrix of the Finite Fourier Transform, a concept intimately related to Fourier series.

Recommended supplementary reading for this section: M.-D. Choi, Tricks or treats with the Hilbert matrix, *American Mathematical Monthly*, Volume 90, May 1983, pp. 301–312; and R. Bhatia, Pinching, trimming, truncating and averaging of matrices, *American Mathematical Monthly*, Volume 107, August–September 2000, pp. 602–608.

A Note on Normalisation

One peculiar feature of writings on Fourier analysis is that the factor 2π is placed at different places by different authors. While no essential feature of the theory changes because of it, this is often confusing. You should note that we have used the following notations:

1. The Fourier coefficients of f are defined as

$$\hat{f}(n) = \frac{1}{2\pi} \int_{-\pi}^{\pi} f(\theta) e^{-in\theta} \, d\theta.$$

2. The convolution of f and g is defined as

$$(f * g)(x) = \int_{-\pi}^{\pi} f(x - t) g(t) \, dt.$$

3. The Dirichlet kernel is defined as

$$D_N(t) = \frac{1}{2\pi} \sum_{n=-N}^{N} e^{int}.$$

4. For functions on a bounded interval I we have defined L_2 and L_1 norms as

$$\|f\|_2 = \left(\frac{1}{|I|} \int_I |f(x)|^2 \, dx \right)^{1/2},$$

$$\|f\|_1 = \frac{1}{|I|} \int_I |f(x)| \, dx,$$

where $|I|$ is the length of the interval I.

When you read other books you should check whether the same conventions are being followed. For example, some authors may not put in the factor $1/|I|$ while defining norms; others may put in the factor $1/2\pi$ while defining $f * g$. Then some of their theorems will look different because the factor 2π if suppressed at one place will appear elsewhere. See the remark following Theorem 4.2.4.

As a consolation we might note that in spite of several international conferences on "units and nomenclatures" many engineering texts use—sometimes on the same page— the f.p.s., the c.g.s., the m.k.s. and the rationalised m.k.s. systems of units.

A Brief Bibliography

This is a brief bibliography. Some of the books listed here provide the necessary background. Others are suggested as collateral and further reading.

Analysis

Several generations of mathematics students have learnt their analysis from the classic

W. Rudin, *Principles of Mathematical Analysis,* McGraw Hill, first published in 1953, 3rd ed., 1976.

Chapters 1–8 of Rudin provide adequate preparation for reading most of this book. For some of the sections more advanced facts about integration are needed. These may be found in Chapter 10 of Rudin, and in greater detail in Part 1 of another famous text:

H. L. Royden, *Real Analysis,* MacMillan, 3rd ed., 1988.

Elementary facts about differential equations that we have used in this book are generally taught as applications of the Calculus. These may be found, for example, in

R. Courant, *Differential and Integral Calculus,* 2 volumes, Wiley Classics Library Edition, 1988.

or, in

W. Kaplan, *Advanced Calculus,* Pearson Addison-Wesley, 5th ed., 2002.

For a more detailed study of differential equations the reader may see

W. E. Boyce and R. C. DiPrima, *Elementary Differential Equations and Boundary Value Problems,* 7th ed., Wiley Text Books, 2002.

Chapter 10 of this book, titled "Partial differential equations and Fourier Series," contains topics close to the ones we have studied in the early sections.

We have used elementary properties of complex numbers and functions in this book. At places we have pointed out connections with deeper facts in Complex Analysis. The standard reference for this is

L. V. Ahlfors, *Complex Analysis,* 3rd ed., McGraw Hill, 1978.

Fourier series

Two well-known books on Fourier series, available in inexpensive editions, are

H.S. Carslaw, *Introduction to the theory of Fourier's Series and Integrals,* Dover Publications, 1952,

and

G.H. Hardy and W.W. Rogosinski, *Fourier Series*, Cambridge University Press, 1944.

Less well known and, unfortunately, not easily available is the excellent little book

R.T. Seeley, *An Introduction to Fourier Series and Integrals*, W.A. Benjamin Co., 1966.

Our beginning sections are greatly influenced by, and closely follow, Seeley's approach. A more recent book which is very pleasant reading is

T.W. Körner, *Fourier Analysis*, Cambridge University Press, paperback edition, 1989.

All these books are at an intermediate level. More advanced books, for which a thorough knowledge of functional analysis is essential include

H. Helson, *Harmonic Analysis,* Hindustan Book Agency, 1995,

Y. Katznelson, *An Introduction to Harmonic Analysis*, Dover Publications, 1976,

and

H. Dym and H.P. McKean, *Fourier Series and Integrals*, Academic Press, 1972.

General reading

We have emphasized the history of Fourier Series and their connections with other subjects. The following books were mentioned in the text.

W. Dunham, *Journey Through Genius,* Penguin Books, 1991.

W. Dunham, *Euler: The Master of Us All,* Mathematical Association of America, 1999.

V. M. Tikhomirov, *Stories about Maxima and Minima,* American Mathematical Society, 1990.

E. Maor, *Trigonometric Delights,* Princeton University Press, 1998.

J. Arndt and C. Haenel, π *Unleashed,* Springer, 2001.

A. Weil, *Number Theory: An Approach Through History from Hammurapi to Legendre,* Birkhäuser, 1984.

History and biography

The recent book

J.-P. Kahane and P.-G. Lemarié-Rieuseet, *Fourier Series and Wavelets,* Gordon and Breach, 1995

provides a scholarly history of Fourier Analysis and also a treatment of more recent topics.
Fourier's original book *Théorie Analytique de la Chaleur* has been translated into English as *The Analytical Theory of Heat* published in a paperback edition by Dover. It is regarded as both a scientific and a literary masterpiece. A biography of Fourier is contained in the famous book *Men of Mathematics* by E.T. Bell, Simon and Schuster, 1937. A scholarly full length biography is *Joseph Fourier, The Man and the Physicist* by J. Herivell, Oxford, 1975. A few facts about Fourier may induce you to look at these books. He was a revolutionary, accused of being a "terrorist" and was almost killed on that charge; he was a soldier in Napoleon's army that invaded Egypt and later a successful governor of a province in France; he wrote a survey of Egyptian history that has been acknowledged as an outstanding work by historians; he was the Director of the Statistical Bureau of the Seine and is known among demographers for his role in developing governmental statistics in France; his book on heat propagation has been repeatedly published as a part of "Great Books" collections by several publishers. A most striking counterexample to the common belief that a scientist is someone forever bent over his equipment or papers, oblivious of and indifferent to the world around.

... if man wishes to know the aspect of the heavens at successive epochs separated by a great number of centuries, if the actions of gravity and of heat are exerted in the interior of the earth at depths which will always be inaccessible, mathematical analysis can yet lay hold of the laws of these phenomena. It makes them present and measurable, and seems to be a faculty of the human mind destined to supplement the shortness of life and the imperfection of the senses; and what is still more remarkable, it follows the same course in the study of all phenomena; it interprets them by the same language, as if to attest the unity and simplicity of the plan of the universe, and to make still more evident that unchangeable order which presides over all natural causes.

Joseph Fourier

Index

Notation

About the Author

Rajendra Bhatia received his PhD from the Indian Statistical Institute in New Delhi, India. He has held visiting appointments at various universities across the world—Sapporo, Kuwait, Ljubljana, Pisa, Bielefeld, Lisbon, Toronto, and Berkeley, among others.

He is a member of the Indian Mathematical Society, the MAA, the AMS, and the International Linear Algebra Society (on whose board of directors he has served). He has been on the editorial boards of several mathematics journals such as *Linear Algebra and Its Applications*, *SIAM Journal on Matrix Analysis*, and *Linear and Multilinear Algebra*. He has been a recipient of the Indian National Science Academy Medal for Young Scientists and Shanti Swarup Bhatnagar Prize of the Indian Council for Scientific and Industrial Research. He is a Fellow of the Indian Academy of Sciences.

His previous widely acclaimed books include: *Perturbation Bounds for Matrix Eigenvalues*, and *Matrix Analysis*.

Structural Wonders

Golden Gate Bridge

Judy Wearing and Tom Riddolls

W

Weigl Publishers Inc.

Published by Weigl Publishers Inc.
350 5th Avenue, Suite 3304, PMB 6G
New York, NY 10118-0069

Website: www.weigl.com

Library of Congress Cataloging-in-Publication Data

Library of Congress Cataloging-in-Publication Data available upon request.
Fax 1-866-44-WEIGL for the attention of the Publishing Records department.

ISBN 978-1-60596-136-1 (hard cover)
ISBN 978-1-60596-137-8 (soft cover)

Printed in China
1 2 3 4 5 6 7 8 9 0 13 12 11 10 09

Photograph Credits
Every reasonable effort has been made to trace ownership and to obtain
permission to reprint copyright material. The publishers would be pleased
to have any errors or omissions brought to their attention so that they may
be corrected in subsequent printings.

Weigl acknowledges Getty Images as its primary image supplier for this title.

Page 24 right: Henrik Sendelbach: page 27 East Bridge; Photo by Rutahsa Adventures,
www.rutahsa.com: page 24 right; Alex Needham: page 27 Xihoumen Bridge.

All of the internet URLs given in the book were valid at the time of publication.
However, due to the dynamic nature of the Internet, some addresses may have
changed, or sites may have ceased to exist since publication. While the author
and publisher regret any inconvenience this may cause readers, no responsibility
for any such changes can be accepted by either the author or the publisher.

Project Coordinators: Heather C. Hudak, Heather Kissock
Design: Terry Paulhus

Contents

What is the Golden Gate Bridge?

Spanning the mouth of San Francisco Bay, the Golden Gate Bridge is one of the United States' best-known landmarks. The bridge was built in the 1930s to connect Marin County to the city of San Franciso. Prior to its construction, people had to take a **ferry** across the bay or travel around the entire edge of it to get from one place to the other. That trip was more than 100 miles (161 kilometers) long. Building the bridge reduced that distance to 1.3 miles (2.1 km), and the trip took far less time.

The Golden Gate Bridge faced many challenges before and during its construction. It was built during the Great Depression, a time when the economy was slow and many people were without work. Even though building the bridge would provide people with jobs, it was difficult to raise the money needed for the project. There were also many people who felt that a bridge could not be built across the bay due to its high winds, strong ocean currents, and almost-constant, heavy fog.

At least one person, however, believed that the bridge could be built. An **engineer** by the name of Joseph Strauss drew a plan to build the bridge and gave it to local authorities. In spite of people's concerns, he was given approval to begin the project. It took more than 10 years to organize the construction of the bridge, but it was built in only four years.

When it opened in 1937, the Golden Gate Bridge was the longest **suspension bridge** in the world. That record stood for 27 years. Today, the bridge is known for its commanding towers, sweeping cables, and colorful orange frame.

Quick Bites

- The Golden Gate Bridge was named for the water it spans across. The mouth of the San Francisco Bay is actually a **strait** called the Golden Gate. Explorer John Charles Frémont gave it this name in 1846. It reminded him of the Golden Horn, a piece of land that forms the harbor of Istanbul in Turkey.
- The Golden Gate Bridge is a toll bridge. It costs money for a vehicle to take the trip across.

Building History

San Francisco Bay was a busy place in the early 20th century. The ferries that crossed the bay were always bustling, and the ports were often crowded with lineups of cars. A local engineer began asking bridge engineers across the United States if it was possible to build a bridge across the open ocean of the Golden Gate. Many said it would be impossible or that it would cost $100 million, which was too much to pay for the project. Bridge engineer Joseph Strauss said it was very possible to build such a bridge and that it would only cost $30 million.

The project began to take shape with the creation of the Golden Gate Bridge and Highway District. Formed in 1928, its goal was to organize the financing, design, and construction of the bridge. The District, along with Joseph Strauss, believed the bridge should be built, but they had to fight to make it happen.

It took workers 30 to 40 minutes each morning to climb up to the site to work.

At first, many organizations were against the idea. The ferry companies did not want a bridge because it would ruin their business. The U.S. Department of War, which owned the land the bridge would join, was concerned that the bridge would interfere with ships coming into the bay. Over time, these hurdles were overcome, and the Department of War gave permission for the bridge to be built on its land.

The Golden Gate uses the longest bridge cables ever made. They could circle the Earth three times.

TIMELINE OF CONSTRUCTION

1928: The Golden Gate Bridge and Highway District is formed.

1929: Joseph Strauss is made chief engineer.

1932: Bonds are issued to fund construction.

1933: Construction begins. The two **anchorages** and the **foundation** for the north tower are completed.

1935: The south tower is finished.

1936: The main cables and steel structure for the road are ready.

1937: The road is added. On May 27th, the Golden Gate Bridge is opened for people to walk across. Vehicles are allowed on the bridge the next day.

1938: Joseph Strauss dies of a heart attack not long after the bridge is complete.

A suspension bridge is a kind of "hanging road."

However, there still was no money for building. Due to the state of the economy at the time, the Golden Gate Bridge and Highway District found it difficult to raise the funds. In 1930, it asked the district's voters for permission to issue $35 million in **bonds**. Voters risked their houses and farms to provide the **collateral** needed for the bonds. They believed that tolls collected from traffic on the bridge would pay back the money. Over time, they were proven right.

Construction of the bridge began in 1933. Factories across the country made 83,000 tons (75,000 tonnes) of steel parts and shipped them to San Francisco. Construction workers accomplished many things no one else in the world had ever done. They built a tower under water in the open ocean and strung gigantic cables 220 feet (67 meters) above the water.

Four years later, on May 27, 1937, the Golden Gate Bridge was finished and opened to traffic. Strauss had built the bridge on time.

Each cable on the Golden Gate Bridge is made up of 61 highly compressed strands.

Due to its unique harp-like design, when the bridge opened in 1937, *The San Francisco Chronicle* referred to it as a "thirty-five million dollar steel harp."

Big Ideas

Although Joseph Strauss is credited as the designer of the Golden Gate Bridge, he received a great deal of help from other bridge engineers and designers. They worked cooperatively to build a bridge that would be grand in appearance but also able to withstand the physical environment in which it would be located.

The first design that Strauss submitted was for a **cantilever**-suspension bridge. When engineer Leon S. Moisseiff and **architect** Irving F. Morrow became involved in the project, it was decided that a suspension bridge would work better and look nicer. With the construction style confirmed, the design of the bridge could be planned. Irving F. Morrow chose to create a bridge that had an Art Deco style. Art Deco is a type of design known for its geometric patterns, bright colors, and elegant, curving surfaces.

The finished product is a symbol of this popular 1920s and 1930s design style. The curves are apparent in the sweeping cables that climb from each end of the bridge to the two rectangular towers. Wide, upright ribbing on each tower's horizontal **struts** gives the bridge texture when sunlight shines on it.

Web Link:
To find out more about suspension bridges , go to www.pbs.org/wgbh/nova/bridge/meetsusp.html.

1) With its graceful curves and tall towers, the Golden Gate Bridge is one of the most spectacular bridges in the country. 2) About 40 million cars cross the Golden Gate Bridge each year. 3) The Golden Gate Bridge's orange paint provides a splash of color that is part of the Art Deco style.

Profile:
Joseph Strauss

Joseph Baermann Strauss was born in Cincinnati, Ohio, on January 9, 1870. His mother and father were of German background and had artistic leanings. His father, Raphael Strauss, was a writer and painter, who was known for his paintings of U.S. presidents. His mother was a pianist. Joseph acquired his father's love for the written word and wrote poetry in his spare time. In fact, when he graduated from the University of Cincinnati in 1892, he was the class president and the class poet as well. His main course of study, however, was in commerce and economics.

Strauss' fascination with bridges came as a result of a hospital stay while attending the University of Cincinnati. At 5 feet, 3 inches (1.6 m) tall, Strauss was not a typical size for a football player. Still, he decided to try out for the university's football team. Strauss was injured during the tryouts and was sent to the hospital. From his room, he could see the Cincinnati-Covington Bridge, one of America's first suspension bridges. This bridge inspired him to become a bridge engineer.

Strauss moved to Chicago, Illinois, where he went to work as an **apprentice draftsman** for a local ironworks company. He later joined the engineering firm of Ralph Modjeski, a well-known Polish-American bridge engineer. Strauss worked as Modjeski's principal assistant.

THE WORK OF JOSEPH STRAUSS

Burnside Bridge, Portland, Oregon (1926)
The Burnside Bridge crosses the Willamette River in Portland, Oregon. It is a **bascule drawbridge** designed by Joseph Strauss and built in 1926. Opening and closing the bridge to let boats pass through takes about eight minutes.

Cherry Street Bridge, Toronto, Ontario, Canada (1931)
Strauss's bridge company built the Cherry Street Bridge in Toronto, Ontario, in 1931. It allows Cherry Street to cross over the Toronto ship harbor channel. The north end of the bridge has a 750-ton (680-tonne) **counterweight** that helps lift the center span.

In 1955, a statue of Joseph Strauss was placed on a landing near the bridge in his honor.

While there, Strauss became interested in designing moveable bridges. He invented two types of bridge. One was a type of bascule drawbridge, which uses counterweights to balance the moveable parts. The other was a type of vertical-lift bridge. On this type of bridge, the moveable part is lifted straight up, or vertically, from the rest of the bridge in a flat, or horizontal, position.

With his reputation as an engineer growing, Strauss left Modjeski's firm and formed his own company. Bascule bridges were the company's specialty, and it built 400 of these bridges across North America. However, they were fairly small in size. Joseph had always dreamed of building a big bridge. When he heard about the Golden Gate Bridge project, he saw an opportunity to achieve this goal. He fought for more than 10 years to make the bridge a reality.

The effort it took to build the Golden Gate Bridge tired Joseph. He died in Los Angeles, California, on May 16, 1938, a year after the bridge was completed. A statue of him now stands on the San Francisco side of the bridge.

Jackson Boulevard Bridge, Chicago (1916)
Chicago's Jackson Boulevard Bridge was built in 1916. A bascule bridge, it extends over the south branch of the Chicago River. The bridge is still in use today. About 8,200 vehicles travel across it every day.

Lewis and Clark Bridge, Washington (1930)
The Lewis and Clark Bridge extends across the Columbia River, joining the towns of Longview, Washington and Rainier, Oregon. When it was built, it was the longest cantilever bridge in the United States. Originally called the Longview Bridge, it was renamed the Lewis and Clark Bridge in 1980 after the historic explorers.

The Science Behind the Building

A bridge must support its own weight as well as the weight of any objects that will travel over it. The materials used to make up the bridge, as well as the design of the bridge, all must work together to create a strong structure. The Golden Gate Bridge is elegant in its look, but it is also very strong. As a result, it is able to withstand high winds and fast-moving waters, as well as hold the traffic that crosses it each day.

The Science of Suspension

Suspension bridges, such as the Golden Gate Bridge, work due to the forces of **compression** and **tension**. These forces are used to support the weight of the bridge and its cargo. They do this by allowing the weight of the bridge and its traffic to be transferred to other points.

The roadway of the Golden Gate Bridge hangs from cables that stretch across the bridge's entire span, running through the bridge's two towers. The towers help to support the roadway's weight. Compression presses down on the roadway, but the cables counter this pressure by pushing it

Fog makes it difficult for incoming ships to see the Golden Gate Bridge. To make it more visible, the U.S. Navy wanted the bridge to be painted black with yellow stripes.

into the towers. From there, the pressure moves down the towers and into the earth. The bridge stays stable as a result.

The forces of tension are handled by the bridge's supporting cables. These cables are attached to the bridge's anchorages. The weight of the bridge, along with the weight of the traffic crossing the bridge, stretch the cables tight, causing tension. This tension is relayed to the anchorages, which, like the towers, are built into the ground. The tension from the cables is directed to the ground, taking the pressure off the bridge itself and keeping it firmly in place.

Trusses

Due to their length and use of cables, suspension bridges are exposed to a force called torsion. This is a twisting force normally caused by high winds. If not controlled, it can seriously damage a bridge. On the Golden Gate Bridge, this force is combatted through the use of trusses. A truss is a type of framework made up of metal or wood that gives a structure **rigidity**. The trusses on the Golden Gate Bridge provide reinforcement to the bridge's **deck**. This makes it more resistant to twisting forces. In recent years, the trusses have been refitted with newer, stronger materials. This helps to keep the bridge from collapsing due to twisting movements during earthquakes.

Web Link:
To find out more about how compression and tension affect bridges, go to http://science. howstuffworks.com/bridge.htm

Science and Technology

Building the Golden Gate Bridge required planning, labor, and the best technology available at the time. Sometimes, technology relies on scientific principles from the past. In the case of the Golden Gate Bridge, the construction machines used were often made up of **simple machines**, such as pulleys and wedges. However, for some of the construction, new methods and machinery were used.

Pulleys

A very large amount of steel had to be moved during the building of the Golden Gate Bridge. Due to the weight of the steel, a crane was used to move the pieces into place. Cranes use pulleys to lift heavy loads easily. Pulleys are wheels with a groove around the edge. Ropes or cables are pulled along the groove to reduce the weight of the load. The more wheels the pulley system has, the more the weight is shared between them.

Rivets

There are more than one million steel rivets holding the Golden Gate Bridge together. A rivet is a short shaft of metal with a rounded head that is narrow at one end and gradually widens at the other end. When two pieces of steel are joined, a hole is drilled through both. The rivet is put through the hole and is then hit with a hammer. The hammer spreads the end of the rivet. This makes one end of the rivet wider, holding the two pieces of metal in place.

The pulleys used to construct the Golden Gate Bridge had many wheels. This made it easier to lift pieces of the bridge that weighed many tons.

The flattened end of a rivet holds the steel pieces firmly together.

Spinning the Cable

At the time, no other object had ever been built that was similar to the suspension cables of the Golden Gate Bridge. Each cable is made of small wires that have been bound together very tightly. Great pressure is needed to force thousands of tons of steel wire into a perfect round shape. To do this, the wires were bound around giant spools that were attached to the bridge's anchorages. One end of the wire was tied to part of the anchorage, while the rest of the wire was wound around the giant wheel. This wheel was then pushed across the bridge, with the cable unwinding as the wheel rolled. Once the wheel reached the other side of the bridge, the end of the cable was attached to that side's anchorage, and the process began again. Spinning the cables took a great deal of time. To speed up the process, two wheels were used at the same time. Each started at opposite ends of the bridge and passed by each other in the center.

It took 80,000 miles (128,748 km) of wire to make the Golden Gate Bridge's main cables.

Quick Bites

- From July to October, foghorns on the bridge can be heard for about five hours each day. They help boats avoid a crash with the towers and the shore.
- There are 600,000 rivets in each tower that hold the steel pieces together. Every one of them was tightened by hand.

Computer-Aided Design

Architects are trained professionals who work with clients to design structures. Before anything is built, they make detailed drawings or models. These plans are important tools that help people visualize what the structure will look like. A blueprint is a detailed diagram that shows where all the parts of the structure will be placed. Walls, doors, windows, plumbing, electrical wiring, and other details are mapped out on the blueprint. Blueprints act as a guide for engineers and builders during construction.

For centuries, architects and builders worked without the aid of computers. Sketches and blueprints were drawn by hand. Highly skilled drafters would draw very technical designs. Today, this process is done using computers and sophisticated software programs. Architects use CAD, or computer-aided design, throughout the design process. Early CAD systems used computers to draft building plans. Today's computer programs can do much more. They can build three-dimensional models and computer simulations of how a building will look. They can also calculate the effects of different physical forces on the structure. Using CAD, today's architects can build more complex structures at lower cost and in less time.

Computer-aided design programs have been used since the 1960s.

Eye on Design

Testing the Old with the New

Today, engineers use highly technical computer programs to make drawings and calculations.

When the Golden Gate Bridge was built, engineers did most of their planning with pencils and paper. Today, engineers can use computer programs to help them plan and figure out complex problems. Computers also can be used to research any questions an engineer might have about bridge construction, such as how much a bridge can sway without breaking, or what can happen to a bridge during an earthquake.

Some companies are hired to study old bridges. They use computer programs to look for weak places on a bridge. All kinds of information about the bridge are put into the computer. Measurements about how the bridge moves in the wind, and the size and strength of its different parts are compiled. In this way, a computer model of the bridge is made. The model is put through a series of tests to see how it will react to certain situations, such as earthquakes and other forces. The computer calculates where the structure might fail.

This kind of testing has been done on the Golden Gate Bridge. Engineers found ways the bridge could be strengthened to help it survive earthquakes. Work is continually being done to correct these problems.

MEASURING THE GOLDEN GATE BRIDGE

Location

The Golden Gate Bridge is located at the mouth of San Francisco Bay, in the state of California, on the west coast of the United States.

Height

- The towers stretch 746 feet (227 m) above the water.
- The deck is 220 feet (67 m) above the water.

Length

- The bridge is 8,981 feet (2,737 m) long in total. The longest span, from the north to south tower, is 4,200 feet (1,280 m) long.
- The bridge's main cables are 7,650 feet (2,332 m) long.

Weight

- Each anchorage weighs 60,000 tons (54,431 tonnes).
- The entire bridge weighs 887,000 tons (804,673 tonnes).

Other Interesting Facts

- On opening day in 1937, 18,000 people lined up at 6:00 a.m. to be the first to cross the bridge. Some even crossed on stilts or walked backwards.
- During construction, a net was strung under the bridge. It saved the lives of 19 workers and is now a standard safety feature in bridge building.

Movement

The Golden Gate Bridge has been designed to sway 27.7 feet (8.4 m) in the event of an earthquake.

Environmental Viewpoint

Tides enter and leave the bay twice a day, crashing against the bridge's underwater supports. Winds whip past the cables and push against the road deck, causing the bridge to sway. The salty sea water and high humidity get under the bridge's paint and cause metal **corrosion**. As a result, the bridge's condition needs to be checked all the time so that any potential problems can be fixed.

In some cases, minor repairs are done on the bridge. The paint is "touched up" in areas where breaks have occurred. Steelworkers replace rivets and steel bars that are badly corroded. Sometimes, however, the bridge needs more serious work done to it. The bridge has undergone many renovations since it was first built. In the 1950s, it was outfitted with a bracing system to improve its stability. In the 1980s, the entire deck was replaced due to salt and humidity damage. This work took years to complete.

In addition to these problems, San Francisco Bay lies in a region where two of the Earth's **tectonic plates** meet. When these plates move against each other, an earthquake can occur. The San Francisco area is under constant threat of earthquakes. When one occurs, the bridge needs to be able to withstand the intense movement. It has already survived one earthquake, in 1989, but concerns remain about the bridge's strength and stability.

Storms sometimes threaten the safety of vehicles driving across the Golden Gate Bridge. However, the bridge has only been closed three times due to high winds.

In 1997, work began to make the Golden Gate Bridge more resistant to earthquakes. Almost every part of the bridge is having repairs or renovations done to it. Foundations and trusses are being strengthened. Supporting towers and braces are being replaced. New parts are being added to the bridge to help it adjust to the extreme shifting an earthquake can cause. The project is taking place in three stages. It is expected to be complete in 2013.

BRIDGE POLLUTION

While the environment has a direct impact on the Golden Gate Bridge, the bridge has also affected its environment. With the high volume of vehicles that cross the bridge everyday comes pollution caused by car exhaust. This exhaust puts dangerous gases into the air. They can cause problems ranging from difficulty in breathing to **acid rain**.

Actions have been taken to reduce this pollution problem. Campaigns have been run that urge people to use city buses instead of their cars to travel across the bridge. This means more people can travel with less exhaust created. Those who do travel in cars are encouraged to carpool. People going to work with two or more passengers in their car do not have to pay the $6 bridge toll.

Golden Gate Bridge   **21**

Construction Careers

The Golden Gate Bridge was built during the Great Depression. This was a time when many people did not have jobs. No one knows how many people worked on the bridge, but they came from all kinds of backgrounds. Farmers and office clerks became steelworkers so that they could help construct the bridge. Cement workers built the anchorages. Engineers planned, designed, and managed the construction process.

Cement Masons

Cement masons specialize in handling cement and concrete. They pour wet concrete into molds and make sure it spreads to the necessary thickness. They then level and smooth the surfaces and edges of the cement. Throughout the process, cement masons check how the wind and temperature are affecting the concrete and fix any potential problems. Working with cement is very physical work. Cement masons often kneel and bend over the surface they are working on. They also carry heavy bags of cement. This requires a great deal of physical strength.

Steelworkers

Steel makes up a good part of the Golden Gate Bridge. Steelworkers put the metal pieces together. They read the engineer's blueprints, and then, put the bridge's framework together. This involved welding or bolting the correct pieces in place. Steelworkers continue to work on the Golden Gate Bridge. They need to be in good physical condition and have good mechanical skills. Steelworkers should be able to use a variety of tools to inspect the bridge parts and replace any that are worn.

Painters

The Golden Gate Bridge painters have to be very comfortable with extreme heights. When painting, they are often suspended high on the bridge. Here, they feel the full brunt of high winds, fog, rain, and other weather conditions while wearing only a small harness to hold them in place. The Golden Gate Bridge needs paint to protect its metal from rusting in the salty sea air, so the role of the painter is very important. The entire bridge has only been repainted three times. Most of the time, painters touch up problem areas. They find spots that are degrading as a result of smog and car exhaust. They quickly remove the damaged paint using hand tools or sandblasters. After cleaning the area, they apply a fresh coat of paint. It takes 38 painters, working year-round, to maintain the paint on the bridge.

Structural Engineers

Structural engineers make sure structures are built to be safe, strong, and stable. They ensure that the bridge's design and construction materials will survive the pressures of the bridge's environment. In the case of the Golden Gate Bridge, this includes strong water currents, high winds, and potential earthquakes.

Structural engineering requires strong technical knowledge and a solid understanding of the scientific principles involved in construction processes. Most structural engineers have at least a bachelor's degree in engineering from a college or university.

Web Link:
To find out more about structural engineering, visit www.pbs.org/wgbh/buildingbig/profile/career/structural.html.

Notable Structures

Suspension bridges have been used for a very long time. Ancient peoples, such as the Incas, built bridges suspended by cables. Over time, these bridges became more complex. They were designed to span longer and wider areas and to hold heavier cargo.

Akashi Kaikyō Bridge

Built: 1998

Location: Kobe and Awaji-shima, Japan

Design: Honshu-Shikoku Bridge Authority

Description: Also known as the Pearl Bridge, the Akashi Kaikyo Bridge is now the longest suspension bridge in the world. It is 12,828 feet (3,910 m) long, with towers that are 928 feet (283 m) tall. This bridge is built to stand up to strong winds, **tsunamis**, and earthquakes.

Inca Suspension Bridge

Built: 16th century to present

Location: Huinchiri, Peru

Design: Incas

Description: Known as the last Inca hanging bridge, this bridge extends across the Apurimac River. It is made from stiff grasses woven into a strong, thick rope that is 150 feet (46 m) long. Villagers from both sides of the river come together for three days every year to make new rope for it. They do this to honor their past.

These bridges often became known for their beauty. With their high towers, gliding cables, and elegant appearance, suspension bridges often become important landmarks for the places where they reside. Some countries have even placed them in national protection programs to preserve them for years to come.

Brooklyn Bridge

Built: 1883

Location: Manhattan and Brooklyn, New York, United States

Design: John Roebling

Description: The Brooklyn Bridge joins Brooklyn to New York City. It spans 1,595 feet (486 m) and is the second-busiest bridge in New York City. During its construction, John Roebling died. His son, Washington, took over the project but became ill. His wife helped him complete the bridge.

Union Bridge

Built: 1820

Location: Horncliffe, United Kingdom

Design: Captain Samuel Brown

Description: One lane of traffic travels 449 feet (137 m) over the River Tweed on the Union Bridge. As the oldest suspension bridge still in use, it is protected as an ancient monument. This means that it is considered to be a structure of national importance and, as a result, cannot be changed in any way without special permission.

Suspension Bridges Around the World

Suspension bridges can be found all over the world. Some are used for car traffic, while others carry pedestrians and trains.

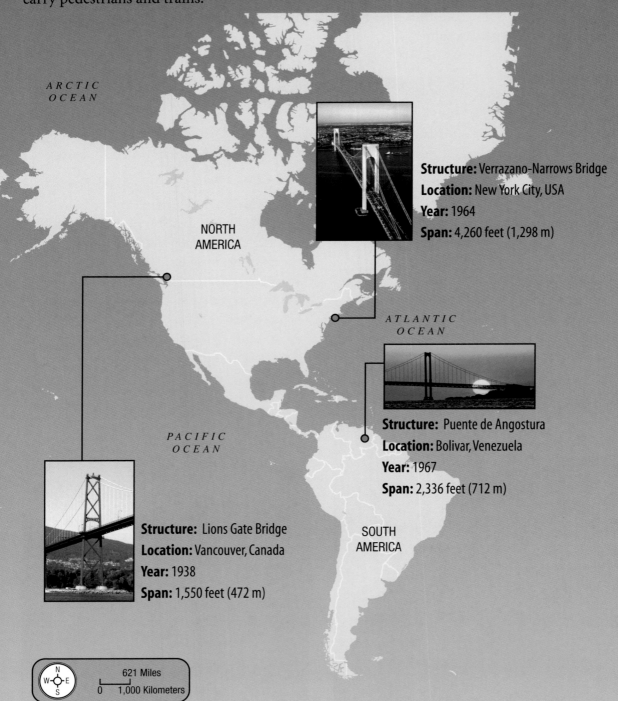

Structure: Verrazano-Narrows Bridge
Location: New York City, USA
Year: 1964
Span: 4,260 feet (1,298 m)

Structure: Puente de Angostura
Location: Bolivar, Venezuela
Year: 1967
Span: 2,336 feet (712 m)

Structure: Lions Gate Bridge
Location: Vancouver, Canada
Year: 1938
Span: 1,550 feet (472 m)

ARCTIC OCEAN

NORTH AMERICA

ATLANTIC OCEAN

PACIFIC OCEAN

SOUTH AMERICA

621 Miles
0   1,000 Kilometers

When there are long distances to cover, suspension bridges are the choice of engineers. This map shows some of the world's better known suspension bridges.

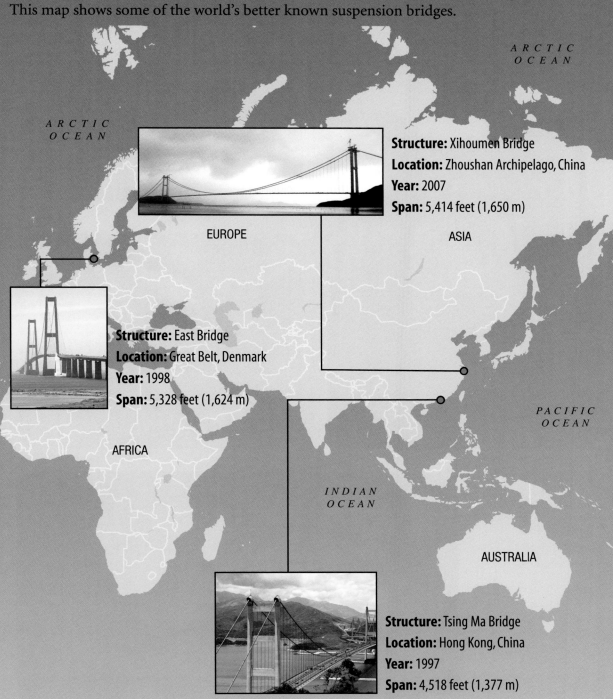

Structure: Xihoumen Bridge
Location: Zhoushan Archipelago, China
Year: 2007
Span: 5,414 feet (1,650 m)

Structure: East Bridge
Location: Great Belt, Denmark
Year: 1998
Span: 5,328 feet (1,624 m)

Structure: Tsing Ma Bridge
Location: Hong Kong, China
Year: 1997
Span: 4,518 feet (1,377 m)

ARCTIC OCEAN

ARCTIC OCEAN

EUROPE

ASIA

AFRICA

PACIFIC OCEAN

INDIAN OCEAN

AUSTRALIA

Quiz

Q Was the Golden Gate the first suspension bridge in the world?

A No. Suspension bridges, in some form, have been around for centuries. Ancient cultures, such as the Incas, were known to have built suspension bridges.

Q What is the name of the Golden Gate Bridge's design style?

A The bridge was designed using the Art Deco style. This was a popular design style in the 1920s.

Q Did Joseph Strauss design the bridge by himself?

A No. Many engineers and designers helped Joseph Strauss plan the Golden Gate Bridge.

Q What are the two main forces suspension bridges rely on?

A Suspension bridges work as a result of the forces of compression and tension. These forces support the weight of the bridge and its traffic by diverting the pressure to other points.

Build a Suspension Bridge

A suspension bridge has five main parts. These are towers, anchorages, main cables, suspending cables, and a deck. Try building a model suspension bridge using materials you may have around the house.

Materials
- six paper towel rolls
- glue
- long pipe cleaners
- scissors
- cardboard pizza box
- toy car
- corrugated cardboard
- pen or pencil

Instructions
1. Cut two of the paper towel rolls in half. The full cardboard rolls will be your bridge's towers, and the half-rolls will be its anchorages.
2. Glue the towers to the center of the pizza box. Then, glue two anchorages to each edge of the box. Let the glue dry.
3. Cut small slits into the tops of the towers and anchorages. To begin making the bridge's cable, make a knot in the end of a pipe cleaner. Put the knotted end into one of the slits. Link other pipe cleaners to the end to make the cable longer. String the pipe cleaners from one set of anchorages to the two sets of towers and then to the other set of anchorages.
4. To make the bridge deck, measure the width and length of the bridge. Draw these measurements onto a piece of corrugated cardboard, and then cut out the shape.
5. To suspend the bridge's deck, take several pipe cleaners and bend their tips around the bridge's main cables. Make sure the pipe cleaners extend along the length of the entire bridge, from anchorage to anchorage. Pass these pipe cleaners under the deck and then attach them to the cables on the other side of the bridge. If you need more length, twist more pipe cleaners together. When complete, adjust the pipe cleaners until the deck hangs parallel to the base.
6. How much weight does your bridge hold? Try running a toy car across it. Does the bridge hold the weight of the car? If not, how could you make your bridge stronger?

Further Research

You can find more information on the Golden Gate Bridge, suspension bridges around the world, or the building of bridges at your local library or on the Internet.

Websites

For more information about the Golden Gate Bridge, visit www.goldengatebridge.org

Learn more about suspension bridges at www.bridgemeister.com

Find out more about how bridges work, visit www.howstuffworks.com/bridge.htm

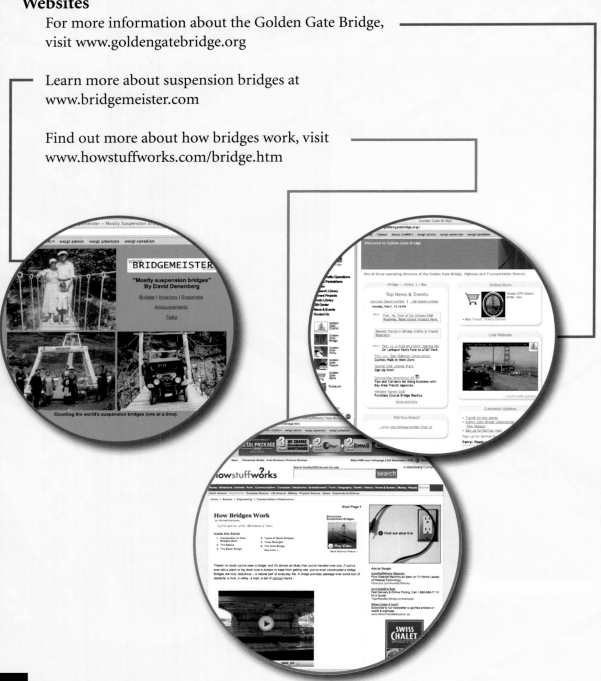

Glossary

acid rain: rain that has been contaminated by air pollution; can damage buildings, harm wildlife, and pollute water supplies

anchorages: structures used to hold something else firmly in place

apprentice: someone who is being trained for a job

architect: someone who designs and supervises the construction of a building or other structure

bascule drawbridge: a bridge that opens to allow boats to pass through; uses a heavy counterweight to help open and close the bridge

bonds: certificates of debt issued in order to raise funds

cantilever: a structural framework that is fixed at one end only

collateral: security pledged for the repayment of a loan

compression: the act of being flattened or squeezed together by pressure

corrosion: a process in which an object is worn away by a chemical action

counterweight: an object that balances opposing forces

deck: a platform on which bridge traffic travels

draftsman: someone who draws or sketches building plans

engineer: someone who applies scientific principles to the design of structures

ferry: a boat used to carry people and vehicles across water

foundation: the part of a structure that helps support its weight

rigidity: stiffness

simple machines: devices that overcome resistance at one point by applying force at some other point

strait: a narrow passage of water that links two larger bodies of water

struts: the bracing pieces of a structure's framework

suspension bridge: a bridge that has a deck supported by cables that are anchored at both ends

tectonic plates: pieces of Earth's surface that are constantly moving

tension: the state of being stretched

tsunamis: high sea waves caused by natural events, such as earthquakes

Index